Safety Supervision
Second Edition

Dan Petersen

AMERICAN SOCIETY OF SAFETY ENGINEERS ❖❖❖ Des Plaines, Illinois

Copyright © 1999 by Dan Petersen

All rights reserved under International and Pan-American Copyright Conventions. No reproduction in any form or by any means, electronic or mechanical, including photocopying, is authorized without permission in writing from the publisher. All inquires should be addressed to: Manager Technical Publications, American Society of Safety Engineers, 1800 E. Oakton Street, Des Plaines, IL 60018-2187.

Managing Editor: Michael F. Burditt
Page Design and Compostion: William M. Johnson
Cover Design: Publication Design, Inc.

Library of Congress Cataloging-in-Publication Data

Petersen, Dan.
    Safety supervision / Dan Petersen. -- 2nd ed.
      p.   cm.
    Includes index.
    ISBN 1-885581-22-X (hardcover. : alk. paper)
    1. Industrial safety.   2. Supervision of employees.   I. Title.
HD7261.P48   1999
658.3'82--dc21

98-49110
CIP

> This book is designed to provide accurate and authoritative information in regard to the subject matter covered. It is furnished with the understanding that the author is not hereby engaged in rendering legal or other professional services. If legal advice or other professional assistance is required, the services of a competent professional person should be sought.

Manufactured in the United States of America
Second Edition

10 9 8 7 6 5 4 3

Contents

Preface v

1 Introduction
Where Are You Now? 2
The Executive 4
The Middle Manager 5
The Supervisor 7
The Worker 7

2 Background
Causation Theories 9
Employee Training 10
Supervisory Training 10
Records 10
Safety Media 11
Today's Safety Philosophy 13
The Real Formula 18
Where Do We Go From Here? 18

3 Change
External Environment 21
Internal Environment (Within an Organization) 22
Total Quality Management (TQM) Concepts 27
Behavior-Based Safety Management 30
Why Is Change Necessary? 31

4 Drivers
The Performance Model 34
Accountability Systems 37
Roles 44

5 The Supervisor's Job
What is the Supervisory Job? 47
Your Responsibilities As A Supervisor 50
What Authority Do You Have? 52
Your Authority As A Supervisor 55
How Are You Measured? 57
Your Accountability As A Supervisor 59
Where Does Safety Fit Into Your Job? 59

6 The Safety Job
How Safety Gets Accomplished 65
The Supervisor's Role in Safety 70
Safety Where There Is No Supervisor 71
Our Attempts To Date 74
Successful Teams 75
An Example 76

7 Key Safety Tasks
Investigating for Causes 83
Costs of Accidents 90
Inspecting for Hazards 93
Coaching 101
Motivating 104

8 The Dynamics of Supervision
Understanding Workers 107
The Sins of Supervision 107
Incongruency Theory 108
What Turns Them On 110
Dual Factor Theory 112
Attitudes 113
The Group 120
Discipline 128
The Problem Worker 134

9 Safety Techniques That Work: Investigating for Accident Cause
Safety Sampling (SS) 141
Statistical Safety Control (SSC) 145
Technique of Operations Review (TOR) 148
The TOR System 148
Incident Recall Technique (IRT) 152

10 Safety Techniques That Work: Inspecting For Hazards
Job Safety Analysis   157
Hazard Hunt (HH)   161
OSHA Compliance Check   163
Ergonomic Analysis   169

11 Safety Techniques That Work: Coaching
Job Safety Observation (JSO)   175
One-on-One Contacts   178
Safe Behavior Reinforcement   179
Stress Assessment Technique   180

12 Safety Techniques That Work: Motivating
Worker Safety Analysis   185
Inverse Performance Appraisals   189
Safety Improvement Teams (SIT's)   189
Climate Analysis   191

Appendices
Supervisory Self-Appraisal   195
Stress Tests   207
Safety and Health Law   219

Index   229

Preface to the Second Edition

Many years ago I wrote *Safety Supervision* for the American Management Association (AMACOM). When I did that in 1976, the world of safety was a quite different place. It was still largely before OSHA's influence became pervasive, and it was in an era where we believed that the supervisor was the "Key Man" in safety, as Heinrich taught us.

Everything has changed since then. OSHA is here and the supervisor is no longer perceived as the "Key Man" in safety. In some organizations there are no supervisors, or they may be referred to as "Team Leaders" or "Facilitators." In some the so-called supervisor is not a member of management, but is really a member of the union. (What is his/her role?)

So the whole concept of "safety supervision" appears to have changed. Or has it? Do workers today have no supervision—no bosses? In some cases, yes: in some companies team management has led to high performance operations. And it usually works very well. But most of the time it is not quite that simple or pure. Supervisors are now team leaders—a mere title change— since decisions are still made above. In some cases there is pure chaos—nobody is running the show. And, in fact, in most cases there has been little change—workers still are supervised by bosses as they always were. For all of our wonderful talk about participation and involvement, in the majority of organizations there still exists a boss, or a boss structure—from top to bottom—from CEO to worker, and in that structure there is always a crucial link—the worker to his or her direct boss. By whatever name, that person is a supervisor.

This book is about that person—that one person who connects the hourly employee to the organization—the supervisor. In the Army we called him a sergeant, and he was the only person that made it all work. In industry we used to call him a foreman—and he got the same results—he made it all work.

In the Army I was a lieutenant—one of the most worthless positions—without the sergeants I would have been ridiculous; at least I think I knew it at the time, and they (the sergeants) saved my _____ over and over. In industry the first line supervisor makes the organization run when the organization allows them to. When the organization uses that position only as a training ground, with constant churn, the only hope is they have good enough employees to run the operation in spite of the situation.

This book is dedicated to two first line supervisors that I knew well, and who were just as typical as the ones who saved me in the Army:

- My dad, JP, in the Omaha packing houses;
- My father-in-law, AB, at Northwestern Bell (then a part of the AT&T system).

In 1996, VNR published three revisions of books I had previously written: *Human Error Reduction & Safety Management*; *Safety by Objectives—What Gets Measured and Rewarded Gets Done*; and *Analyzing Safety System Effectiveness*. As I was revising these, I also was revising this book, *Safety Supervision*. We postponed publishing this revision—three was enough for one year. However, we referred to *Safety Supervision* several times in the other revisions, particularly in *Safety by Objectives* and *Analyzing Safety System Effectiveness*, since Supervison was intended to be a companion piece to the other books. The later two revisions were aimed at management—how to set up systems to hold people accountable for their performance—for doing things regularly, proactively, to control losses.

I am constantly asked in seminars after discussing accountability, "Accountable to do what?" This book describes some options, and what supervisors might actually do which will result in fewer accidents and injuries. In the other books accountability systems are discussed, this book suggests some actions that supervisors (or teams) might be held accountable to do.

Dan Petersen

CHAPTER 1 Introduction

It has been a fundamental belief since the beginning of industrial safety activity that if anything is to be accomplished to promote safety, it will be accomplished by the supervisor. Safety professionals believe this—so do managers at all levels. Unfortunately, having this belief has not gotten the job done.

Over the years there has been substantial progress in making industrial conditions safer, but most of the real progress occurred during the early years of the effort. In recent years, progress has ceased.

I believe one of the causes of this lack of progress is our inability to transform our fundamental belief about supervisory responsibility into supervisory action. Most supervisors in American industry do not do a good job in safety. If they did, the job would be getting done. The principal reasons for this are:

—Management has seldom been sufficiently clear about what it wants done in safety. The prevailing attitude is: "It's your responsibility, now go do something about it."

—Management almost never follows up to see if the supervisor has in fact done anything in safety. At most, someone may look at a safety record.

—Management almost always measures supervisors in areas other than safety to ensure that performance standards are being met.

Given this set of circumstances, why should a supervisor spend any time and effort on safety? The aim of this book is to change this situation, first by clarifying and simplifying what supervisors should do in order to achieve safety. There is no mystery to it—as long as certain things are done regularly and routinely, results should come. Secondly, I hope to show how management can clearly and simply measure performance by supervisors in safety.

This book is not intended to be a comprehensive reference for the supervisor in safety. It will not answer specific questions about what is dangerous and what is safe. There are probably references available somewhere in every company to do this. The National Safety Council's *Accident Prevention Manual for Business and Industry* is one such reference that is present in most companies.

WHERE ARE YOU NOW?

Before continuing, perhaps you ought to determine where you stand now in safety. You'll get a better idea by studying the 15 questions that follow and analyzing your responses. For each question choose the answer that best describes your current situation.

1. **Do I have a job description that indicates precisely what I am to do in regard to safety?**
   (a) No
   (b) I have a job description, but it doesn't mention safety.
   (c) Yes

2. **Do I know exactly how much authority I have in safety?**
   (a) No
   (b) Yes
   (c) I've discussed it with my boss, and we've come to some agreements.

3. **Do I know exactly how I'm going to be measured in safety?**
   (a) No
   (b) Yes
   (c) I've discussed it with my boss, and we've come to some agreements.

4. **Do I know exactly what I am expected to do in safety?**
   (a) No
   (b) All that counts are results—no accidents.
   (c) Yes

5. **Do I know what is considered acceptable performance in safety?**
   (a) No
   (b) No accidents, I guess.
   (c) Yes

6. **How much time do I spend on safety?**
   (a) No time
   (b) It's a constant job.
   (c) A few hours a week

7. **Have I read the OSHA standards?**
   (a) What's that?
   (b) Yes
   (c) Those sections that apply to me.

8. **Have I made a list of all violations of the law in my area?**
   (a) No
   (b) Once
   (c) Regularly

9. **Have I set up a system of priorities?**
    (a) No
    (b) Yes
    (c) Yes. It's in operation.
10. **Have I kept my boss aware of my status in OSHA compliance?**
    (a) Heaven forbid
    (b) Once I did.
    (c) Regularly
11. **Have I documented everything I've done in safety and in OSHA preparation?**
    (a) No
    (b) In part
    (c) Yes
12. **Do I know what turns my people off?**
    (a) Who cares?
    (b) Everything, I think.
    (c) Yes, I know.
13. **Do I know what turns them on?**
    (a) Who cares?
    (b) Nothing
    (c) Yes, I know.
14. **Do I know when to use discipline?**
    (a) Always
    (b) Never
    (c) Yes, I know.
15. **Do I know how to work with problem people?**
    (a) The same way as with anyone else.
    (b) No, I don't.
    (c) Yes, I do.

There are, obviously, many other questions that could be asked, but these will give you an idea of the type of material that follows.

Score yourself three points for any (c) answer, two points for a (b) answer, and one point for an (a) answer. If you scored 45, you have no worries and probably no accidents. If you scored between 35 and 44, you need to work at it still. If you scored 25 to 34, you're not quite there, and if you scored under 24, maybe you ought to see the boss about a transfer—or read on.

This book is written to the supervisor—that person in the organization that makes it happen, whether it be safety, quality, production, or anything else that management wants done.

But we've learned over the years that in many (probably most) organizations, safety just doesn't get accomplished by the supervisors (or by anyone else) in the organization. It is talked about, but not acted on; it is important when you're not busy, but quickly disappears in your rush periods. This is all pretty easy to understand when we look at management theory and research about why supervisors do what they do—and why they don't do what management "says" it wants done.

Therefore, in this book about supervisor performance in safety, before we even start addressing the supervisors, we first address those that have influence over what the supervisor does every day.

THE EXECUTIVE

If supervisors are not performing (that is doing what they "ought" to be doing, and getting the results you want), first of all you, the executive, should blame yourself. The supervisor tries, as best he can, to give you what you want. Usually, there are some problems in doing that. For instance:

- He may not know what it is you want him to do.
- He may be getting conflicting signals from you. ("Safety First"—but get the rush orders out today!)
- He may judge what you want by the way that you measure him instead of by what you say.
- He may judge what you want by how you reward him instead of by what you say.
- There may not be enough hours in the day to do all of the things that you say you want done.
- It could be that your systems have loaded him with so much record keeping and report writings (computer time) that he cannot do what you want.
- It could be that your "rightsizing" has spread the ranks of your supervisors so thin that they simply cannot supervise.
- It could be that, in going to teams you've gotten rid of supervisors and not replaced them, and not thought out team roles, team functions, team responsibilities, or team accountabilities.
- It could be you expect your people to do things beyond their capabilities.
- etc.

These are just a few examples of conditions that exist today in many organizations. Downsizing (by whatever name you call it), re-engineering, and a raft of other terms used to shake-up and change organizations without clear role definitions for whomever is left—without adequate training for those left in new settings, new jobs, new titles, etc.—often leaves only chaos.

I have found in my consulting that following the introduction of new approaches from the executive level, the new approaches are not clearly perceived at the very next level, are quite fuzzy at the second level down, are not only very fuzzy below that, but are resented, and are hated at the supervisory level, who usually simply do not understand the executive intent.

At the worker level, they know whether or not doing the job safely is important to their boss, and to the organization. Frequently at this level there is no perceived impact in executive-planned changes, but the change never quite gets to their level. At the bottom (where the work is done), whether or not they have a supervisor, a team leader, a facilitator or whatever else he is called, is frankly often immaterial. They just do their work (thank God), and get the job done. The culture ("What goes on around here every day") tells them whether safety is a "core value" or a management lie.

Executives don't have to be irrelevant to the worker—as they often are. Executives can have a powerful impact not only by saying things, but also (more importantly) by making things happen.

In this book are spelled out a number of things a supervisor (by whatever title you use) can do that will get results in safety. At the executive level, all you have to do is decide which accountability style fits you (SCRAPE, SBO, or MENU), as described in Chapter 4, and implement whichever you feel most comfortable with.

This, of course, assumes that (a) you are not satisfied with your current safety results, and (b) you are willing to take an active role to make changes occur. There are plenty of reasons to change today (see Chapters 2 and 3). But it will take you, the executive, to champion the changes needed.

THE MIDDLE MANAGER

If the supervisors are really going to spend their time, effort, etc., on safety to get the results that your executives want, they will only do that if you make it happen. The old textbooks said the supervisor was the key person in safety. That was not true, for we know today that the supervisor reacts to what is important to his boss, you—the middle manager. What is of crucial importance to the middle manager is of crucial importance to the supervisor. He reacts to your goals, desires, wishes, and measures.

The roles in safety are simple and clear-cut:

1. **The role of the first-line supervisor is to carry out some agreed-upon tasks to an acceptable level of performance.**
2. **The roles of middle and upper management are to:**
   a. Ensure subordinate performance.
   b. Ensure the quality of that performance.
   c. Personally engage in some agreed-upon tasks.
3. **The role of the executive is to visibly demonstrate the priority of safety.**
4. **The role of the safety staff is to advise and assist each of the above.**

Obviously, your role as middle manager is the key, for you ensure supervisory performance that ensures worker behavior. As we will discuss later in this book:

The supervisor's role as defined above is relatively singular and simple—to carry out the tasks agreed upon. What are those tasks? While it may depend upon the organization, the tasks might fall into these categories:

Traditional Tasks	**Nontraditional Tasks**
Inspect	Give positive strokes
Hold meetings	Ensure employee participation
Perform one-on-ones	Do worker safety analyses
Investigate accidents	Do force field analyses
Do job safety analyses	Assess climate and priorities
Make observations	Perform crisis intervention
Enforce rules	
Keep records	

In addition to the above, supervisors no doubt will be responsible for certain day-to-day actions not easily spelled out or measured (following standard operation procedures [SOPs]). How well do they understand their responsibilities? They often simply do not know what is expected of them, particularly in safety. Almost always they have no idea of the extent of their authority, and usually they are unclear as to how their performance is being measured, again particularly in safety.

As mentioned earlier, the role of the supervisor is to engage regularly (daily) in some pre-defined tasks. What are those tasks?

That's up to you. This book suggests what they might be.

THE SUPERVISOR

The rest of this book is directed at you, the supervisor. If executives and middle mangers are going to require specific activities from you, what is it that you are going to do to satisfy your safety responsibility? The following spells this out.

Hopefully you will have some flexibility in deciding which tasks and activities you agree to carry out. If you do, the workers will see that you mean it—that safety is important. But it all starts with you showing the way by doing those things every day.

THE WORKER

Your "safety role" is simple—work in such a manner that you won't get hurt, and get involved in your organization's safety activities. And, if you are in a high-performance organization with no supervisor, then your role is much, much larger. You and your peers must take over—do all of the things that must be done to ensure safety. These might be: inspecting, investigation, coaching, motivating, etc., all the things discussed in the remainder of this book.

As a team member it is up to you to make it happen: to create a culture that says that safety is a key value in this organization.

CHAPTER 2 # Background

IN THE EARLY YEARS OF THE SAFETY MOVEMENT, management concentrated entirely on correcting the hazardous physical conditions that existed. This effort showed remarkable results during the first 20 years. In deaths alone, the reduction was from an estimated 18,000 to 21,000 lives lost in 1912 to about 14,500 in 1933. The death rate (deaths per million worker-hours worked) for that period would show an even better reduction. This reduction came largely from cleaning up working conditions. Cleaning up physical conditions came first—possibly because they were so obviously bad, and possibly because people believed that these conditions were actually the cause of injuries.

We have been involved in industrial safety in an organized fashion since 1911. Originally our emphasis was toward improving physical conditions. Since the 1930s we have also considered the unsafe acts of people. Our progress was excellent until the 1960s. Since then we seem to have been slowly losing ground in our battle to control accidents on the job.

Since the 1930s, we have built safety programs, built our techniques and our tools, on a set of principles that were first developed in 1930 and espoused by H. W. Heinrich in his text. He called his principles the "Axioms of Industrial Safety." By looking at five of the most common areas in our safety programs, we can quickly see how our programs were built upon his principles—and how they are changing.

CAUSATION THEORIES

Accident investigations and inspections are integral to all safety programs. OSHA depends almost entirely upon them. This approach is based on a Heinrich axiom that says:

The occurrence of an injury invariably results from a completed sequence of factors, the last one of these being the accident itself. The accident is, in turn, invariably caused or permitted directly by the unsafe act of a person and/or a mechanical or physical hazard. This is commonly known as the "Domino Theory of Accident Causation." It suggests that if we remove the act and/or the conditions (that is, the middle domino), the accident and the injury will not occur.

Newer safety theory disputes this domino theory and replaces it with a multiple causation theory. The multiple causation theory states that accidents are caused by the combination of a number of things, all wrong, that combine at one point in time and result in an injury. This theory suggests that the act, the condition, and the accident itself are all symptoms of something wrong in the management system. The role of safety is not to remove the symptom, but to find our what is wrong with the system.

EMPLOYEE TRAINING

A second area is that of employee training. Our safety programs invariably include aspects such as safety orientation, employee training, and foreman safety talks. These are based upon a Heinrich axiom that states, "The unsafe acts of persons are responsible for a majority of accidents." The underlying assumption of this axiom is that the employees do not know the difference between right and wrong, between safe and unsafe. We know that employees do know, in many cases, what is safe behavior and what is not. In such cases, training is not the solution to the problem.

SUPERVISORY TRAINING

A third area is supervisory training. Safety programs are almost always heavy in the area of supervisory training. This is based on a Heinrich axiom that says, "The supervisor or foreman is the key man in accident prevention." One of the underlying assumptions here is that merely identifying the supervisor as the key man magically makes that supervisor do something about safety. He will not. A second assumption is that the foreman automatically will do something about safety if he is made responsible for safety, and if he knows what to do (that is, he is trained). We now know that this assumption also is wrong.*

Most supervisors today know that they are responsible for safety, and they know what they should be doing, yet they do not do it. Why? Because they usually are not held accountable. That is, they are not measured in safety. Higher management's behavior often says that they either do not know or do not care if supervisors do anything about accidents. Therefore, why should foremen or front-line supervisors do much to prevent accidents? The "key man" axiom must be amended to recognize that "higher management holds the key chain."

RECORDS

Fourth, our safety programs include record keeping and the analysis of these records. The belief is that if we find the most common kinds of incidents and attack them, automatically our costs will come down. This thinking is based upon a Heinrich axiom that states, "The severity of an injury is largely fortuitous." We have interpreted this to mean that the causes of frequency and of severity are the same. Today we know that this is wrong. Severity is predictable in certain situations and under certain circumstances.

SAFETY MEDIA

Fifth, our safety programs include the use of all kinds of safety media, such as posters, banners, films, and literature. This is based on a Heinrich axiom that says, "One of the four basic motives or reasons for the occurrence of unsafe acts is improper attitude."

Do posters, banners, and the like do any good? We really do not know. To be more accurate, so far we know just a little bit about the answer to this question. Research tells us that in some cases these kinds of safety media help, in some cases they do harm, and in some there is absolutely no effect.

The above five areas fairly well describe the state of the art in safety programming. We do many things in our programs, almost all of which are based on the thinking of H. W. Heinrich in 1931.

Safety programs in some companies seem to have just grown: a training program added one year, safety sampling another, and so on. But a closer analysis reveals reasons behind the selection of safety program components.

1. **Knowledge and preference of the safety director.** Most corporate safety directors do the necessary developmental work and sell top management on the elements of a program. It is inconceivable that a safety director would install something he does not believe will work, or something he is not familiar with. Therefore, the corporate safety program is, first of all, a reflection of the personality and knowledge of the safety director. A technically sound safety professional with an inspection orientation leans to the physical categories of safety programs.
2. **Wishes of the boss.** Where top management involves itself in the safety program, or in the decisions on the elements of the program, the manager's personality and style will show through. A good communicator at the top will lean toward a program built heavily on communications and training, for example.
3. **Corporate climate.** The safety program reflects the climate of the organization. The paternal organization may well structure

a safety program heavy on the attitudinal items. A company utilizing an MBO (Management by Objective) concept will tightly fix accountability and be heavy in the behavior change category.

4. **OSHA.** Obviously, OSHA has had a strong influence on safety programs. While the law makes substantial requirements for training and other attitudinal and behavioral activities, most of us concentrate on the physical aspects of OSHA standards. The government's enforcement of the law seems to concentrate on physical aspects too. This changed the look of safety programs: the trend is to reduce emphasis on the attitude influencers and behavior changers, and to put more stress on environment changers. (See Exhibit 2-1.)

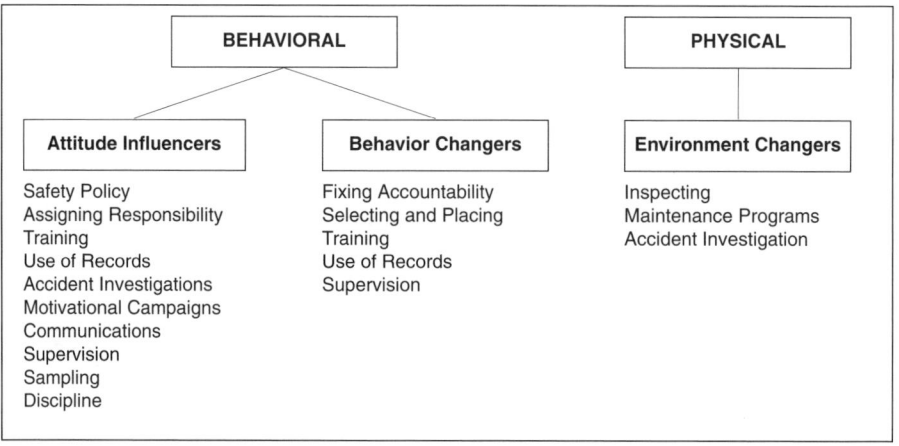

Exhibit 2-1. Common controls in safety programs.

Since OSHA was established, many companies have put the entire emphasis on the physical aspects of safety. Because programs reflect the safety director and the boss, this emphasis probably makes sense. Most managements know about OSHA and would rather not be fined. Most safety directors are (or will be) judged by the number of violations and the size of fines. Under these circumstances, it does not seem logical to place emphasis elsewhere.

As this list of influences indicates, our safety programs today are a composite of what the safety director knows, what the boss likes, what the company likes, and what the federal government wants. Notably absent from the list are the needs of the company and employees.

What does your company need? What do your employees want in a safety program? These are tough questions to answer, so many directors choose simply not to try. It is easier to do what they themselves know, what the boss likes, what fits the company, or what Uncle Sam says. Does this approach achieve the best possible program? Probably not.

Of the above four reasons for safety program content, the fourth by far predominates today. Each year since OSHA came into existence in 1970 there have been more and more regulations at the federal level (not to mention at state and local levels).

Although some of these new regulations may not apply to every organization, many will, and some will be very costly to the organization. To comply with the provisions of the Process Safety Management Standard, the Americans with Disabilities Act, the Hazard Communication Standard, and so on, is not only expensive; it is likely that these costs will never be offset by reduced losses. In actuality these pieces of legislation, and others, may move managerial time and corporate monies away from safety toward compliance. While this book is not about the legal aspects of safety, the reader needs to have a feel for what laws apply in this area. An article explaining this is located in Appendix C.

Rightly or wrongly, this is pretty much the state of the art in safety today. We do a number of things aimed at controlling the physical conditions, or the behavior of people, or aimed at changing the attitudes of employees. We do these things even though they are for the most part based on principles that are dated and questionable in the light of today's thinking. We have done some rethinking of our principles, but our techniques still lag behind them.

TODAY'S SAFETY PHILOSOPHY

In 1971 the first edition of my book *Techniques of Safety Management* was published. It questioned some of the fundamental tenets of safety of the 1960s (which were also the tenets of safety in the 1930s) and brought together a series of basic principles to guide safety efforts in the 1970s. Some of these were restatements or slight changes of Heinrich's axioms, and some were quite different from those axioms.

Here, quite briefly, are the principles outlined in that book:

Principle 1: An unsafe act, or unsafe condition, an accident: all these are symptoms of something wrong in the management system.

We know that behind any accident are many contributing factors. Old theory, however, suggested that we select one of these as the "proximate" cause of the accident, or that we select one unsafe act and/or one unsafe condition. Then we were supposed to remove that condition or act.

This principle suggests not that we narrow our findings to a single factor, but rather that we widen our findings to as many factors as seem applicable. Hence, every accident opens a window through which we can observe the system, the procedures, etc. Also, the theory suggests that besides accidents, other kinds of operational problems result from the same causes. Production tie-ups, problems in quality control, excessive costs, customer complaints, and product failures are caused by the same things as accidents. Eliminating the causes of one organizational problem will eliminate the causes of others.

Principle 2: Certain sets of circumstances can be predicted to produce severe injuries. These circumstances can be identified and controlled.

This principle states that in certain conditions we can predict severity. Therefore, we can attack severity instead of merely hoping to reduce it by attacking frequency.

Principle 3: Safety should be managed like any other company function. Management should direct the safety effort by setting achievable goals, by planning, organizing, and controlling to achieve them.

Perhaps this principle is more important than all the rest. It restates the thought that safety is analogous with quality, cost, and quantity of production. It also goes further and brings the management function into safety (or rather safety into the management function).

Principle 4: The key to effective line safety performance is management procedures that fix accountability.

Any line manager will achieve results in those areas where management is measuring. The concept of "accountability" is important for this measurement.

When line managers are held accountable, they will accept the given responsibility. If they are not held accountable, they will not, in most cases, accept responsibility. They will place their efforts on those things that management is measuring: on production, quality, cost, or wherever the current management pressure is.

Principle 5: The function of safety is to locate and define the operational errors that allow accidents to occur. This function can be carried out in two ways: (1) by asking why—searching for root causes of accidents, and (2) by asking whether or not certain known effective controls are being utilized.

The first part of this principle is borrowed from W. C. Pope and Thomas J. Cresswell, in their article "Safety Programs Management."[1] This article defines safety's function as locating and defining operational errors involving (1) incomplete decision making, (2) faulty judgments, (3) administrative miscalculations, and (4) just plain poor management practices.

The second part of principle 5 suggests that a two-pronged attack is open to us: (1) tracing the symptom (the act, the condition, the accident) back to see why it was allowed to occur, or (2) looking at the system (the procedures) that our company has and asking whether or not certain things are being done in a predetermined manner that is known to be successful.

Principle 6: The causes of unsafe behavior can be identified and classified. Some of the classifications are overload (the improper matching of a person's capacity with the load), traps, and the worker's decision to err. Each cause is one that can be controlled.

Principle 6 suggests that management's task with respect to safety is to identify and deal with the causes of unsafe behavior, not the behavior itself.

Principle 7: In most cases, unsafe behavior is normal behavior; it is the result of normal people reacting to their environment. Management's job is to change the environment that leads to the unsafe behavior.

Principle 7 is an extension of principle 6. It suggests that when people act unsafely, they are not dumb, are not careless, are not children that need to be corrected and changed to make them "right." Rather, it suggests that unsafe behavior is the result of an environment that has been constructed by management. In that environment, it is completely logical and normal to act unsafely.

Principle 8: There are three major subsystems that must be dealt with in building an effective safety system: (1) the physical, (2) the managerial, and (3) the behavioral.

Principle 8 reemphasizes that our task is to change the physical and psychological environment that leads people to unsafe behavior.

Traditionally "safety programs" dealt with the physical environment. Later we looked at management and attempted to build management principles into our safety programs. Today we recognize the need also to look at the behavioral environment—the climate and culture in which the safety system must live.

Principle 9: The safety system should fit the culture of the organization.

The way we manage has changed markedly. And the way we manage safety must also change to be consistent with other functions. To strive for an open and participative culture in an organization and then use a safety program that is directive and authoritarian simply does not work.

Principle 10: There is no one right way to achieve safety in an organization; however, for a safety system to be effective, it must meet certain criteria. The system must:

- Force supervisory performance.
- Involve middle management.
- Have top management visibly showing their commitment.
- Have employee participation.
- Be flexible.
- Be perceived as positive.

These ten principles, restated in Exhibit 2-2, are the underlying theme of this entire book.

For a number of years, probably starting in the mid-1960s or even before, we had observed the need for accountability systems in management as an essential to safety success. Prior to that, safety professionals had pushed for management to be responsible for safety, probably stemming back to the 1930s from one of Heinrich's axioms.

While we have recognized that accountability is the key, most organizations never created truly meaningful accountability systems for safety performance, because they could neither accept nor sell the fact that to achieve accountability requires new measures of performance and new reward structures.

Then came the "Culture Era" and the "Behavior-Based Safety Era." Some organizations were into culture building, (or talking about it), while others were into some variety of behavior molding (or talking about it).

In both of these directions, there has seemed to be an unwritten law that either of these was the answer—nothing else was needed. The "Culture Companies" seemed to jump to a belief that once the new culture is in place, the "accident problem" would be solved—nothing else would be necessary. The new culture posters and pamphlets described the new utopia that now existed. We no doubt needed posters and pamphlets to convince the employees that they were now living in an environment where safety was the key; was a core value. Often management believed things had changed; believed their own writings. Often the employees somehow missed the change—things were the same to them. The "Culture Companies" seemed to feel that since there was a positive safety culture, the old stuff was no longer necessary. Forcing managerial performance through accountability didn't fit anymore—too out of step with the new culture.

> **PRINCIPLES OF SAFETY MANAGEMENT**
>
> 1. An unsafe act, an unsafe condition, an accident: all these are symptoms of something wrong in the management system.
> 2. Certain sets of circumstances can be predicted to produce severe injuries. These circumstances can be identified and controlled:
>
>   > Unusual, nonroutine Nonproductive activities
>   > High energy sources Certain construction situations
>
> 3. Safety should be managed like any other company function. Management should direct the safety effect by setting achievable goals and by planning, organizing, and controlling to achieve them.
> 4. The key to effective line safety performance is management procedures that fix accountability.
> 5. The function of safety is to locate and define the operational errors that allow accidents to occur. This function can be carried out in two ways: (1) by asking why—searching for root causes of accidents, and (2) by asking whether or not certain known effective controls are being utilized.
> 6. The causes of unsafe behavior can be identified and classified. Some of the classifications are overload (improper matching of a person's capacity with the load), traps, and the worker's decision to err. Each cause is one which can be controlled.
> 7. In most cases, unsafe behavior is normal human behavior; it is the result of normal people reacting to their environment. Management's job is to change the environment that leads to the unsafe behavior.
> 8. There are three major subsystems that must be dealt with in building an effective safety system: (1) the physical, (2) the managerial, and (3) the behavioral.
> 9. The safety system should fit the culture of the organization.
> 10. There is no one right way to achieve safety in an organization; however, for a safety system to be effective, it must meet certain criteria. The system must:
>
>   1. *Force supervisory performance.*
>   2. *Involve middle management.*
>   3. *Have top management visibly showing their commitment.*
>   4. *Have employee participation.*
>   5. *Be flexible.*
>   6. *Be perceived as positive.*

Exhibit 2-2. Principles of safety management.

The "Behavior-Based Companies" knew that now that the employees were defining what safe behavior was, and were observing each other, obviously safety was solved, allowing their managers and supervisors or team leaders to concentrate on the truly important things like getting the product out the door, improving quality, doing paperwork, etc. These were the things that they knew to be important, because these were what they were being held accountable to accomplish.

Not all companies went these two routes. Many kept going down their traditional routes—keep the safety program as we've always done it—the easier course. Others opted for traditional programs as outlined by government—an easier and less risky route. Some brought packaged programs. And many did nothing.

Under all of these approaches to safety management, organizations say they have management responsibility, but happily, managers and supervisors or team leaders do not have to *do* anything about safety.

To compound the situation, we have fewer managers and supervisors today; larger spans of control; fewer aides to help; more paperwork to fill out; and, in many cases, supervisors must spend much more time in the office on that great time-saver, the computer.

THE REAL FORMULA

What is the real answer to safety system excellence? The behavior-based people are right—you cannot achieve safety excellence without proper behaviors. But, the culture builders are also right, for you cannot get the proper behaviors without the right culture. And here's the real hook—you cannot get the right culture without accountability. It is not either one or the other; you must build all three, and in the proper order, to achieve excellence in safety.

Management accountability—a system of role definition, correct measures of performance, and adequate rewards contingent upon that performance—forces daily, proactive, managerial and supervisory activities to take place. These actions build a culture that states, "Safety is so important around here that all managers and supervisors have to do something about it every day." When employees believe this, their behaviors will change—both behaviors like working safely, and behaviors like becoming involved in helping the whole process.

Safety excellence can be achieved as long as we re-learn our alphabet:

$$A \longrightarrow C \longrightarrow B \longrightarrow E$$

Accountability builds Culture which gets Behaviors resulting in Excellence.

WHERE DO WE GO FROM HERE?

There is probably no better time than now in most companies to launch a new safety program or build a fire under the present one. Physical conditions are probably at the best level they have ever been. Management is now spending more money in the name of safety than at any other time in its history. Employees today are seeing improvements happening before their eyes, money being spent on their safety as never before. Why not capitalize on this?

Safety programming done in a traditional manner has some very distinct drawbacks and difficulties.

1. Traditionally, safety as a separate function has not been integrated into management's goal setting and achieving process. Safety simply has not been a regular, normal part of the management process. It has been separate and distinct, a program superimposed on the management structure and purpose. Therefore, safety has not necessarily been results-oriented or results-producing.
2. Safety (whatever that might mean to line people) has not been a "known" supervisory skill. Supervisors seldom have known specifically what it is they were to do to achieve "safety," and have spent precious little time on it.
3. Safety historically has had few real measures of performance at the supervisory level. Without measurements, little is ever accomplished.
4. Safety awards (if they have existed at all) have almost never been in any way tied to safety performance for the line manager. Usually if there are rewards, the reward is for good luck more than for actual performance.
5. Safety training historically has involved more preachments than real teaching of skills to achieve results. It has then been almost totally ineffective in terms of getting results for the line manager.

Some of these difficulties have now been or are being overcome with some of our techniques. Others are being overcome with new processes in use in safety management.

Notes

* Although this discussion uses the masculine forms of pronouns, the author's intent is gender inclusiveness, and the reader may substitute "she," "her," etc., as appropriate.
1. W.C. Pope and Thomas J. Cresswell, "Safety Programs Management," *Journal of the American Society of Safety Engineers* (August, 1965).

CHAPTER 3 Change

SAFETY HAS EXPERIENCED A SHIFT IN PARADIGMS, perhaps more than we realize.

A paradigm can be loosely defined as the rules and beliefs that guide our decisions and actions; the "ground rules," if you will, of the game that we are in. When we live within certain parameters, by certain ground rules, we act in certain ways. We are comfortable; we know "what's right" and "what works." When the rules change, we are usually lost for a while. We cannot cope. Nothing works.

First, when you have gone through a paradigm shift, everything you have believed in becomes suspect—you go back to ground zero. So, you first must realize that a paradigm shift has taken place, that the ground rules have changed. This is not always obvious, particularly to those who have spent a lifetime in safety. But changes have happened.

EXTERNAL ENVIRONMENT

- Workers' compensation used to protect the company from lawsuits—it does not do so anymore.
- Executives can go to jail today for not dealing with hazards—manslaughter charges, conspiracy charges, willful violations, etc. No executive today (in his or her right mind) can ignore safety in his or her company; but most executives have no idea whether safety is under control, and they usually do not even know what questions to ask.
- OSHA fines. They used to be $300 for a standard violation. Today fines of $3 to $5 million have been levied under the General Duty Clause.
- We have shifted from objective injuries to subjective injuries; from cuts and bruises to back strains, soft tissue injuries, and psychological stress claims. Guarding machines does not help much in reducing frequency rates of many of these injuries.
- More workplace regulations are coming every year.

How do we cope with this external environment? This is what we are trying to determine. Perhaps we need two separate and distinct staffs—one for safety, one for compliance. The two functions usually have little to do with each other.

INTERNAL ENVIRONMENT (WITHIN AN ORGANIZATION)

Also, internal environment shifts occur within our companies. They include:
- Changing styles and beliefs of management. We have gone from classical management in the 1950s to human relations management in the 1960s to situational management in the 1970s to cultural management in the 1980s to downsizing, TQM, SPC, and reengineering in the 1990s—major shifts in philosophy.

During these major philosophical management shifts, safety has gone its own way, ignoring reality:
- Most safety programs remain "classical" in nature—management decides, and people follow the rules.
- We still talk about the foreman as the key person in safety; meanwhile, companies are heading toward having no supervisors. We do not seem to be integrating safety into our changing management philosophies.

We have had a paradigm shift in the external environment and the internal environments of our organizations. But that is not all. We have had a paradigm shift in safety also—as in our beliefs about what causes an accident on the job. We used to believe accidents were caused by unsafe acts and/or unsafe conditions. Today we believe we have to reveal the root causes of unsafe conditions (a defect in the management system) and, more important, the causes of human error.

We have been studying the causes of human error since World War II. That is more than 50 years of research; we ought to know a lot. For instance, people screw up more when overloaded (physically, physiologically, psychologically). They screw up more when their workstations are poorly designed or when management has built in some traps (piece rates, wage incentives, early quits, etc.). It is normal for people to act unsafely when they are adhering to peer group norms and perceive a greater reward for productivity than safety.

When we shift our emphasis away from unsafe acts and conditions, we have had a paradigm shift. This requires us to question most traditional safety tools: inspections, investigations, JSAs, training, and so on. All of these tools search for symptoms, not for causes—like treating a brain tumor with an aspirin.

So how do we deal at the causal level? Do the "Three E's" of safety (Engineering, Education, and Enforcement) work? They are pretty hopelessly out of date unless we change their traditional meanings. Engineering has to be broadened to include ergonomics, the causes of disease, of stress, and so forth. We also must recognize that traditional safety education and enforcement are clearly not always valuable options in affecting behavior, and will *only* be effective if we treat people maturely, using positive reinforcement, and invite their involvement and participation.

In all of this, we end up attacking our fundamental safety beliefs—Heinrich's "Axioms of Industrial Safety."

Those axioms, mostly untrue and never proven, have been our paradigms in safety. They are no longer. We have had a paradigm shift. What are the new principles? Safety people are now searching for them.

I believe in the principles shown in Exhibit 2-2, which are at least mostly research-based. Needless to say, these principles will be replaced by others in the next paradigm shift.

There is more to the safety paradigm shift:

- We know that we cannot use accident statistics to judge performance at almost any level, as they measure mostly luck. We can, however, use perception surveys and behavior sampling.
- Many (perhaps most) audits are invalid measures of system effectiveness.
- Audits also run counter to the philosophy of having a flexible system that allows employee ownership (today's parameters).
- Safety meetings are being replaced by one-on-one contacts.
- We believe that investigating all accidents is a waste of time. Certainly some investigations are necessary (maybe using SPC tools), but not all.
- We believe that the executive should play a more significant role than simply signing policy. MBWA and other techniques can and are being employed by today's CEOs.
- We know that we must use activity measures to judge safety performance at lower organizational levels.
- We know that behavior change comes quicker and lasts longer by using positive reinforcement than with discipline and punishment.
- Many companies (or their workers) are rewriting their safety rule books.
- We know from communication research that repeating is not the key to communication effectiveness. The keys lie in worker perception of management's credibility and in the meaningfulness of the message sent.

- And most important—all the research on safety programming today points to one fact—there are no magic pills or "essential ingredients" in a safety system.

This only sounds chaotic if we refuse to recognize the paradigm shift in safety. If we recognize this shift, it allows us to do almost anything in our safety system. We can experiment and innovate. We can be creative.

OSHA says not to admit the paradigm shift. Its Safety Program Guidelines emphasize methods used before the paradigm shift. You must comply with the law. But there is nothing in the law that says you cannot have a world-class safety system within the new paradigms. You must do both.

Once you recognize that you are now at ground zero, you simply start upward. Start by sorting out your personal philosophy on things like "What's involved in accident causation?"; "What do I believe are the elements best for my company?"; "What is the culture of my company?"; "What will it look like ten years from now?"; "What elements will protect us and get us to where we want to be?" And for the best input, bring in other people to assist your thinking and restructuring—executives, plant managers, middle managers, supervisors, and, most of all, hourly employees.

A number of safety professionals believe that the total effectiveness of safety programming today is seriously reduced by an inability to integrate safety into the regular management systems and by an inability to effectively relate safety performance to corporate goals. Too often safety professionals still only ask for (or hope for) management support. Too often they are in this ineffective position because they have not demonstrated to management that safety *is* a management-controllable cost—controllable by planning, organization, and management direction.

There are some reasons for our position in the eyes of our managers. Consider these as perhaps the main ones:

1. *We are not using (although they have been developed) the needed tools to quantify our objectives or our progress.* This weakness has been discussed in journal articles at length. While some new techniques have been devised and tested, their use is still so limited in comparison to the use of accident statistics that we cannot really report much progress here. Most safety professionals agree that statistics are ineffective as a measure, as a diagnostic tool, or in communicating to management. Yet for many reasons (trade secrets, not taking time to understand new systems, etc.) we as a profession are not improving our

measures of performance, our communications to management, or our diagnostic tools.
2. *We are not relating safety goals to management objectives.* Some companies are succeeding here today, but they are rare. Few companies set achievable safety goals as they plan corporate goals. Few companies have adapted management's planning system to safety planning. Few safety techniques lend themselves to this kind of projection and planning.
3. *We are not coping with group resources in safety.* We know that employees' decisions as to whether or not to produce are controlled by group norms more than by any other single determinant of behavior. This was proved more than 65 years ago in Mayo's Hawthorne studies. Management today is coping with the phenomenon of the group through participative sharing of decisions, improved communications, etc.
4. *We are not tapping the human resources in our organizations.* Modern management is incorporating knowledge from the behavioral sciences into its newer methods and techniques.

Consider, for instance, the following examples of old management knowledge, techniques, and tools that are in use now in companies.

Job Enrichment

Job enrichment is a concept based upon Frederick Herzberg's research[1], which bears out the fact that responsibility, achievement, and recognition are true motivators and should be built back into jobs if we really hope to excite employees about their work.

Herzberg uses the term *motivation* to describe feelings of accomplishment, professional growth, and professional recognition that are experienced in a job that offers sufficient challenge and scope to the worker. He considers apathy and minimum effort the natural results of jobs that offer the worker no more satisfaction than a paycheck and a decent place to work. These hygiene factors may keep a worker from complaining, but they will not make the worker want to work harder or more efficiently. Offering still more hygiene, in the form of prizes or incentive payments, produces only a temporary effect. According to Herzberg, investments in hygiene reach the point of diminishing returns rapidly and do not, therefore, represent a sound motivational strategy.

Employee-Centeredness

The most effective supervisors are those who are "employee-centered," who create an atmosphere in which employees can work and who remove the obstacles in their way, according to research from Rensis Likert.[2]

Likert found that most organizations can be given a numerical rating on a four-point scale that expresses the prevailing management system. This scale runs from System 1, an arbitrary, coercive, highly authoritarian management style that is seldom encountered any longer in its "pure form," through System 4, which is based on teamwork, mutual confidence and trust, and a genuine respect for the individuals who comprise the organization.

One of Likert's more significant findings is that the more closely a company's management style approximates System 4, the more likely it is to have a record of sustained high productivity, good labor relations, and high profitability. Similarly, the more closely an organization approximates System 1, the more likely it is to be inefficient and to suffer from recurrent financial crises. System 4 Management holds high performance goals for itself and for all employees and makes it clear that it expects them to be met. In other words, setting high goals for efficiency and productivity can be best attained by a system of management that is geared to satisfy real human needs.

Interpersonal Relations Improvement

Interpersonal relations training is common today and is effectively (according to Chris Argyris's data)[3] bridging the gap between mature people's needs for independence, sustained interests, equality rather than subordination, etc., and management's historic principles that treat people like children (close supervision, specialization, etc.).

The problems of worker apathy and lack of effort are not simply a matter of individual laziness. Rather they are often healthy reactions of normal people to an unhealthy environment created by common management policies. More specifically, most adults are motivated to be responsible, self-reliant, and independent. These motives are acquired during childhood through the educational system, the family, and communications media such as books, television, and radio. But the typical organization confines most of its employees to roles that provide little opportunity for responsibility, self-reliance, or independence. On the contrary, too many jobs are designed in ways that make minimal demands on the individual's abilities and that place the responsibility for major decisions in the supervisor's hands.

In effect, such jobs create a childlike role for employees and frustrate their normal motivations for a more adult role. Argyris says that the common reaction of withdrawing one's interest from the job—treating it with indifference or even with a certain degree of contempt—is a necessary defensive maneuver that helps individuals preserve their self-respect. Unfortunately, the cost to the organization of these reactions is heavy: minimal output, low quality, and excessive waste.

Job enrichment, employee-centered supervisors, and improvement in interpersonal relations are examples of many of the ways that management is changing in the light of behavioral science knowledge. These and many other areas indicate the marked changes that are and will be taking place in American industry. Safety management has not begun to utilize the knowledge. There is little in the literature on safety techniques that is built on research, new safety principles, or handling today's workers. In general, management is now beginning to recognize that employees can contribute markedly to organizational effectiveness. Systematic techniques are being used to tap these human resources. So far, safety management has not developed comparable systematic techniques.

Safety management could be leading the way. Through employee participation in safety, job enrichment could be started. What better area is there than safety for a supervisor to begin to establish an employee-centered attitude and behavior? Better interpersonal relations can often be attained in safety more easily than in production because of the unity of the purpose.

TOTAL QUALITY MANAGEMENT (TQM) CONCEPTS

In most companies, quality of performance means initial quality of product and increased productivity. Some companies include quality of safety performance at a much later point, often as an afterthought.

Employee involvement is central to the TQM philosophy, which also includes these concepts:

- Building a new organizational culture that embraces safety.
- Using new tools to solve problems.
- Continuously improving the process.
- Using upstream measures to monitor progress.

Perhaps the best description of TQM is captured in Deming's 14 "Obligations of Management." Although the total quality approach differs greatly from traditional safety concepts, the TQM philosophy works well when applied to safety.

If Deming's "Obligations of Management" were rewritten in safety jargon, they might read as follows:

Management Safety Obligations

1. Concentrate on the long-range goal of developing a world-class system, not on short-term annual accident goals.
2. Discard the philosophy of accepting accidents—they are not acceptable.
3. Use statistical techniques to identify the two sources of accidents—the system and human error.
4. Institute more thorough job skills training.
5. Eliminate dependence on accident investigation. Instead, use proactive approaches such as behavioral sampling, fishbone diagrams, flow charts, etc., to reveal system flaws and achieve continuous system improvement.
6. Provide supervisors (and employees) with knowledge of statistical methods (sampling, control charts, etc.), and ensure that these tools are used to identify areas needing additional study.
7. Reduce fear throughout the organization by encouraging all employees to report system defects and help find solutions.
8. Reduce accidents by designing safety into the process. Train research and design personnel in safety concepts.
9. Eliminate the use of slogans, incentives, posters, and gimmicks to encourage safety.
10. Examine work standards to remove accident traps.

Other aspects of TQM are valuable as well. In fact, measures such as the following are necessary in safety:

- Ask employees to define and solve company problems and identify system weaknesses.
- Provide employees with simple tools to solve problems. These include Pareto charts to determine problems, fishbone diagrams to help brainstorm problem causes, flowcharts to observe the system, and scatter diagrams to determine correlations.
- Replace accident-based statistics with other upstream measures (i.e., behavioral sampling).
- Replace accident-based statistics with alternative downstream measures (i.e., employee perception surveys, employee interviews).

The ten safety obligations represent a marked departure from traditional safety beliefs. Under these new corporate obligations:
- Progress is not measured by accident rates.
- Safety becomes a system rather than a program.
- Statistical techniques drive continuous improvement efforts.
- Accident investigations are either reformed or eliminated.
- Safety sampling and statistical process control tools are used.
- Blame for "unsafe acts" is completely eliminated.
- Focus is on improving the system.
- "Whistle-blowers" are encouraged and supported.
- Employee involvement in problem solving and decision making is formalized via corporate procedures.
- Ergonomic well-being is designed into the workplace.
- Safety slogans and gimmicks are eliminated.
- Emphasis is placed on removing system traps that cause human error.

The move toward TQM in safety means refuting many traditional concepts such as:

1. **Irresponsible acts and conditions cause accidents.**
2. **The Three E's of safety—engineering, education, enforcement—are essential to safety programs.**
3. **Low compliance is sufficient.**
4. **The executive role is only to sign policy.**
5. **Management creates safety rules; employees follow them.**

These beliefs should be replaced with the following axioms:

1. **Accidents are caused by a defective management system and a weak safety culture.**
2. **Many methods can be used to shape behavior, not merely the three E's.**
3. **No magic pill can be prescribed. Practitioners must determine which approaches will work best, depending on situational demands.**
4. **Low compliance has limited influence on safety results.**
5. **Executives must provide safety leadership.**
6. **Decisions made at the bottom—by affected employees—are most effective.**

BEHAVIOR-BASED SAFETY MANAGEMENT

In recent years, the safety community has been told that safety's future progress will come from key "behavior-based" approaches. Yet, while many practitioners agree, they are not quite sure what "behavior-based" means.

Since the 1930s, it has been commonly accepted that most accidents are caused by human behavior (unsafe acts) and that, as a result, safety efforts should focus on reducing the probability of unsafe behavior.

Does this mean that most traditional approaches to safety can be called behavior-based approaches? Since the 1970s, regulations and fear of fines have forced practitioners to concentrate on factors such as physical conditions, health, record-keeping, occupational diseases and ergonomics. Defining such efforts as behavior-based is a stretch. However, training awareness campaigns and use of media, which have continued during these years, do target worker behavior.

Over the last 10 years, practitioners have "found" behavior-based safety. Is it the elusive magic pill sought throughout safety's history or simply the "same old stuff" under a different name?

Typically, behavior-based safety is defined as a process of involving workers in defining the ways they are most likely to be injured, seeking their involvement and obtaining their buy-in, and asking them to observe co-workers in order to determine progress in the reduction of unsafe behaviors.

Behavioral Measurement

A true behavior-based approach to safety requires behavioral measurements. For its upstream measure, the process requires a statistically valid observation/sampling process. (Are employees behaving more safely today than last week/month?) To achieve statistical validity, many observations are needed, with consistency over time (one sampler) looking at all unsafe behaviors, not just a select few.

Statistical validity is also required for downstream measures. For most practitioners, this means replacing traditional measures (frequency rates, incidence rates, severity rates) with valid, meaningful measures (such as process improvements achieved and perception surveys).

What is Behavior-Based Safety?

Behavior-based safety is a management system that defines precisely what behaviors are required from each organizational member (from shop worker to CEO); measures whether these behaviors are present; and reinforces desired behavior regularly (daily, hourly).

Behavior-based safety management systems do not turn safety over to any one level of the organization. Certainly, employee involvement is crucial. But eliminating or isolating management from the process is a problematic approach. Management is legally accountable for safety. To abdicate authority for safety, while retaining responsibility and accountability, is tricky at best, risky at worst.

A fundamental tenet holds that a safety system must meet two primary criteria.

1. **Some system of accountability must define roles; ensure that individuals at all levels possess the knowledge to fulfill those roles; measure role fulfillment; and reward behavior based on role fulfillment.**
2. **The system asks for, allows, requires and ensures participation at all levels.**

WHY IS CHANGE NECESSARY?

TQM can seem complicated (although it is not). Since many traditional beliefs must be changed to achieve success, safety practitioners may ask: Why bother? Some answers:

- Traditional safety programs no longer work (if they ever did). Most are not based on fact (scientific research), and they tend to conflict with both management and behavioral research.
- Accident records continue to deteriorate, as companies focus on complying with regulations based on traditional safety beliefs.
- The safety system must be built into an organization's management structure. Historically, safety has been kept separate from regular management. As management has experienced numerous transitions, safety has been suspended in the classical model: management decides, employees follow orders. As a result, safety has become a foreign subject to many managers and executives.
- As management familiarity with safety has decreased, the external environment has increased management's vulnerability to legislation, criminal liability, higher fines, and so on.
- More "injuries" (i.e., cumulative trauma disorders, stress) are compensable today than ever before. Practitioners cannot address these problems via machine-guarding technology.
- The nation's workers' compensation (WC) system is not only in trouble, it is ill—perhaps terminally. WC carriers are withdrawing from some markets, a trend expected to continue.

Notes

1. Frederick Herzberg, *Work and the Nature of Man* (Cleveland: World Publishing, 1966).
2. Rensis Likert, *New Patterns of Management* (New York: McGraw-Hill Book Co., 1961).
3. Chris Argyris, *Personality and the Organization* (New York: Harper & Row, 1957).

CHAPTER 4 Drivers

FOR WELL OVER 50 YEARS we have been preaching the principle of line responsibility in safety work, and yet there are still supervisors today who say, "Safety is the safety director's job," or "If that's a safety problem, take it up with the safety committee." Worse yet, in many companies, when an accident occurs, it goes on the safety specialist's record instead of on the record of the line supervisor in the department where it occurred. Thus, instead of simply preaching that the line has responsibility, we should have been devising procedures to fix such accountability. When a person is held accountable (is measured) by the boss for something, he or she will accept the responsibility for it. If not held accountable, he or she will not accept the responsibility. Effort will be expended in the area where the boss is measuring.

The attitude of the majority of supervisors today lies somewhere between total acceptance and flat rejection of comprehensive accident prevention programs. Most typical is the organization in which line managers do not shirk this responsibility but do not fully accept it either, nor treat it as they would any of their defined production responsibilities. In most cases their "safety hat" is worn far less often than their "production hat," their "quality hat," their "cost control hat," or their "methods improvement hat." In most organizations, safety is not considered as important to the line manager as many, if not most, of the other duties that he or she performs.

On what does a manager's attitude toward safety depend? It depends on abilities, his or her role perception, and effort. All are important, and a manager will not turn in the kind of performance we want unless we take all three into account (see Exhibit 4-1).

Two basic factors determine how much effort a person puts into a job: (1) his or her opinion of the value of the rewards and (2) the connection the person sees between effort and those rewards. This is true of a manger's total job, as well as of any one segment of it, such as safety.

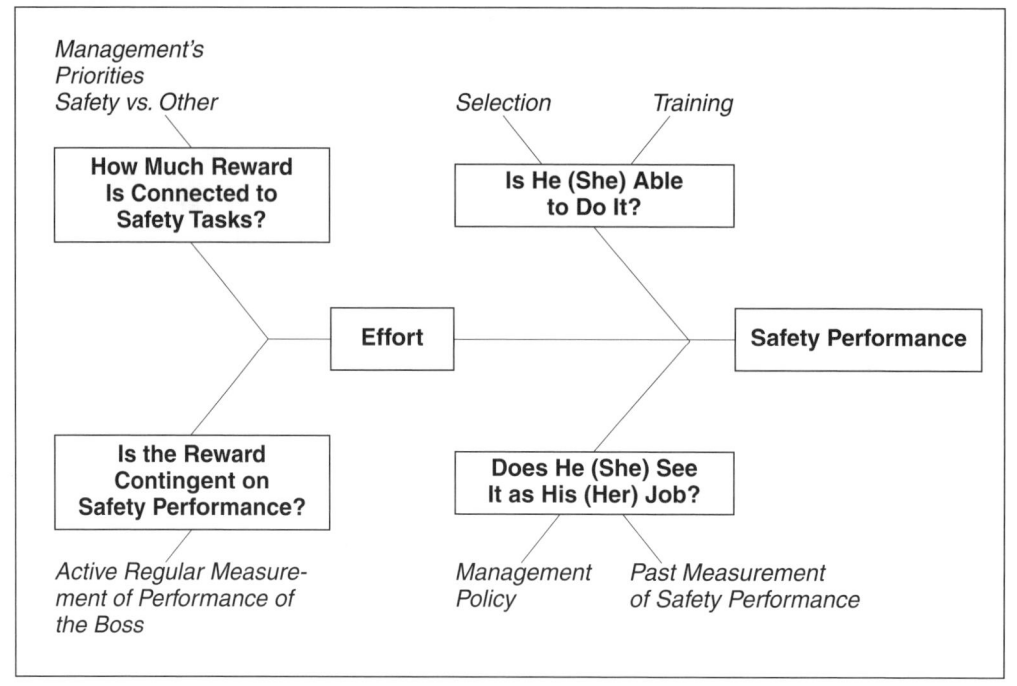

```
                Management's
                Priorities
                Safety vs. Other       Selection      Training
                    ┌──────────────┐              ┌──────────────┐
                    │ How Much Reward│              │ Is He (She) Able│
                    │ Is Connected to│              │   to Do It?    │
                    │  Safety Tasks? │              └──────────────┘
                    └──────────────┘
                              \         ┌────────┐         /           ┌──────────────────┐
                               \────────│ Effort │────────/────────────│ Safety Performance│
                               /        └────────┘        \            └──────────────────┘
                    ┌──────────────┐              ┌──────────────┐
                    │ Is the Reward │              │ Does He (She) See│
                    │ Contingent on │              │ It as His (Her) Job?│
                    │Safety Performance?│          └──────────────┘
                    └──────────────┘
                Active Regular Measure-   Management     Past Measurement
                ment of Performance of    Policy         of Safety Performance
                the Boss
```

Exhibit 4-1. Supervisory safety performance model. *(Adapted from the Lawler and Porter model.)*

THE PERFORMANCE MODEL

The Value of Rewards

Rewards for safety performance are no different from rewards for performance in any other area where management asks for performance from the supervisor. While our model focuses primarily on positive rewards (peer acceptance, subordinate approval, enhancing the likelihood of promotion, merit salary increases, higher bonus, intrinsic feelings of accomplishment, pat-on-the-back, compliment from the boss), it also could mean negative reward (chew-out, lower and harder pat-on-the-back, reprimand).

The main difficulty in developing a reward system is not in determining what the rewards will be for performance but rather in determining when the reward should be given, when the supervisor has "earned" it. For in safety we have precious few good measurement tools that tell us when a supervisor is performing.

The manager looks at the work situation and asks, "What will be my reward if I expend effort and achieve a particular goal?" If the supervisor considers that the value of the reward that management will give for achieving the goal is great enough, he or she will decide to expend the effort.

"Reward" here means much more than just financial reward. It includes all the things that motivate people: recognition, chance for advancement, increased pay. Most research into supervisory motivation today indicates that advancement and responsibility are the two greatest motivators.

In assessing whether rewards really depend upon effort, managers ask the following kinds of questions:

- Will my efforts here actually bring about the results wanted, or are factors involved that are beyond my control? (The latter seems a distinct possibility in safety.)
- Will I actually get that reward if I achieve the goal?
- Will management reward me better for achieving other goals?
- Will it reward the other manager (in promotion) because of seniority, regardless of my performance?
- Is safety really that important to management, or are other areas more crucial to it right now?
- Can management really effectively measure my performance in safety, or can I let it slide a little without management's knowing?
- Can I show better results in safety or in some other area?

The line managers unconsciously ask these questions and others before determining how much effort to expend on safety. They must get the right answers before deciding to make the effort needed for results. Often line managers decide that their personal goals would be better achieved by expending efforts in other areas, and too often their analysis is correct because management is rewarding other areas more than safety.

Changing this situation is the single greatest task of the safety professional. Change can be achieved by instituting better measurement of line safety performance and by offering better rewards for line safety achievement.

Ability

Job performance does not depend simply on the effort that managers expend. It depends also on the abilities they bring to the task, both inherent capabilities and specialized knowledge in the particular field of endeavor. In accident prevention, this means that we must ensure, through training, that line managers have sufficient safety knowledge to control their workers and the conditions under which these people work. In most industries lack of knowledge is not a problem, for line managers usually know far more about safety than they apply. Managers can achieve remarkable results on their accident records merely by applying their management knowledge, even if they have little safety knowledge. If a manger does not have adequate safety knowledge, the problem is easily handled through training.

Role Perception

Role perception is even more important than ability. Line managers' perceptions of their safety role determine the direction in which they will apply their efforts. Lawler and Porter describe a good role perception as one in which the manager's views concerning placement of effort correspond closely with the views of those who will be evaluating his or her performance.

In safety, role perception has to do with whether line managers know what management wants in accident control and with whether they know what their duties are. In evaluating role perception, the safety professional should search for answers to some questions about the organization and about each line manager in it. These questions concern the content and the effectiveness of management's policy on safety, the adequacy of supervisory training, company safety procedures, and the systems used to fix accountability. There are four key considerations:

1. **Is the required performance clearly defined?**
2. **Is the accountability system in place? (Here we are looking at how the person is measured and rewarded.)**
3. **Does each person know how to do what is expected?**
4. **Is the perceived reward enough to capture the person's attention and ensure his or her performance?**

Supervisory Performance

What drives performance is the supervisors' perception of what the boss wants done—their perception of how the boss will measure them, and their perception of how they will be rewarded for that performance. To restate what the research shows, these questions dictate supervisory performance:

- What is the expected action?
- What is the expected reward?
- How are the two connected?
- What is the numbers game? (How is performance measured?)
- How will my actions affect me today and in the future?

ACCOUNTABILITY SYSTEMS

Any accountability system that defines, validly measures, and adequately rewards will work. Here are some examples.

SCRAPE

SCRAPE is a systematic method of measuring accident prevention effort. Most companies measure accountability through analysis of results. Monthly accident reports at most plants suggest that supervisors should be judged by the number and the cost of accidents that occur under their jurisdiction. We should also judge line supervisors by what they do to control losses. SCRAPE is one simple way of doing this. It is as simple as deciding what supervisors are to do and then measuring to see that they do it.

The first step in SCRAPE is to determine specifically what the line managers are to do in safety. Normally this falls into the categories of (1) making physical inspections of the department, (2) training or coaching people, (3) investigating accidents, (4) attending meetings of the workers' boss, (5) establishing safety contacts with the people, and (6) orienting new people. There could be many other tasks, however.

Every week each supervisor will fill out a small form (Exhibit 4-2).

```
Department _____ Week of _____
(1) Inspection made on _____ # corrections _____
(2) 5-minute safety talk on _____ # present _____
(3) # accidents _____ # investigated _____
    Corrections _____
(4) Individual contacts
Names _____
    _____
    _____
(5) Management meeting attended on _____
(6) New person (names):          Oriented on (dates):
    _____      _____
    _____      _____
    _____      _____
```

Exhibit 4-2. SCRAPE system.

Management, on the basis of this form, spot-checks the quality of the work done in all six areas, and rates the accident prevention effort.

With a SCRAPE system, there is tight accountability and little to no flexibility. Each supervisor is required to engage in the same activities, as indicated above. Obviously SCRAPE fits well in a top-down company, with a relatively directive, authoritarian style. It can also be used in the early installation of an accountability system, and flexibility can follow later.

Safety by Objectives

Historically our safety programs have failed in some essential elements. Many programs are far from producing behavior that could be considered goal-directed. Responsibilities, even with written policy, are often unclear. Participation in goal setting and decision making is almost nonexistent. Feedback and reinforcement are slow and often not connected to the amount of effort expended in safety (especially when the number or the severity of accidents is the measuring stick). Planning is minimal, and while results are often measured, freedom of decision or control is seldom left to the lower levels. And, finally, imagination and creativity are rare commodities in most safety programs. The principles of MBO, adapted to safety programming (SBO), can overcome some of these failings (see Exhibit 4-3).

These are the steps of SBO:

1. *Obtain management-supervision agreement (with staff safety consultation) on objectives.* In the installation stages of an SBO program, the agreement will emphasize not only results objectives but also activities objectives. Initially, then, the agreement reached will be on strategies and objectives (what means, tools, and resources to be used, as well as results). Once the program is under way, only objectives are agreed to.
2. *Give each supervisor an opportunity to perform.* Once agreements are completed, leave supervisors alone to proceed with their action plans. Require only progress reports.
3. *Let them know how they are doing.* With quantified objectives (either result or activity objectives must be quantified) give regular, current, and pertinent feedback so they can adjust their plans when they see the need.
4. *Help, guide, and train.* Both management and safety staff fulfill this role. Safety staff provides the technical and safety technical

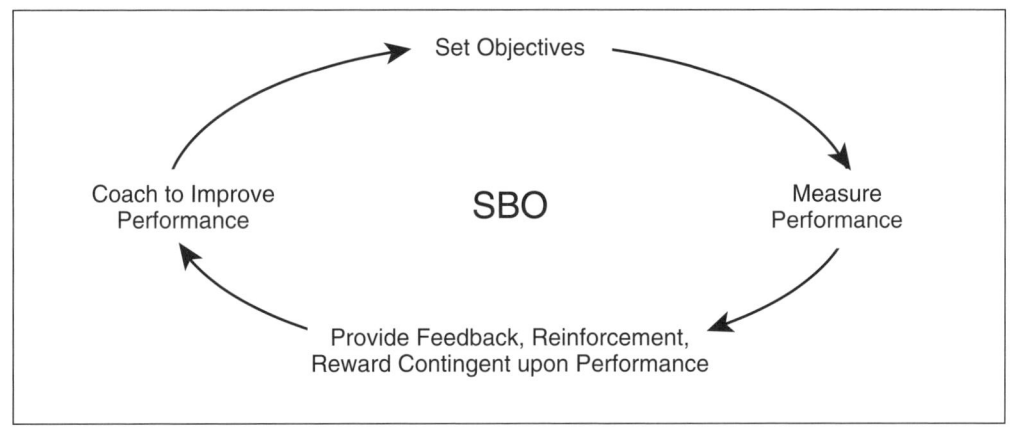

```
              Set Objectives
         ↗                    ↘
Coach to Improve    SBO      Measure
Performance                  Performance
         ↖                    ↙
    Provide Feedback, Reinforcement,
    Reward Contingent upon Performance
```

Exhibit 4-3. SBO System.

expertise, while management provides the managerial help when asked, guiding when indicated, and training at the outset.

5. *Reward them according to their progress.* This requires a reward system that is geared to the progress made toward agreed-upon objectives. The various managerial rewards should be used: pay, status, advancement, recognition.

The SBO approach has been installed in a variety of different industries: brewing (Adolph Coors Company), chemical (DuPont Corp.), railroading (Union Pacific, Santa Fe), paper (Hoerner-Waldorf), and others.

The results, of course, are not uniform. There are some successes and some less successful applications, as much depends on the installation, the commitment of management, and the meaningfulness of the objectives that were set. It has, of course, been installed differently in each organization. Some have left the objective-setting areas wide open—totally up to the supervisors and their bosses. Others have limited supervisory choices to certain allowable strategies. One organization's management spelled out 12 areas that were believed to be areas of slippage in the safety program. Most have required tasks and additional optional ones that a supervisor can select from a menu, then set goals for those selected.

For the most part, SBO works and continues to work once installed. One company reported a 75 percent reduction in the frequency rate in the first three months of the program. One claimed a six-month savings of $2.3 million. One reported a 67 percent reduction in frequency continuing after three years.

With an SBO system, there is tight accountability and almost total flexibility. Each supervisor can select any activities to engage in to satisfy his or her safety responsibility, as long as there is mutual agreement with his or her boss.

> **XYZ COMPANY**
> **ACCOUNTABILITIES--FIRST LINE SUPERVISOR**
>
> **GENERAL**
>
> The key accountability of the First Level Manager is to carry out the tasks defined below.
>
> **TASKS**
>
> Required tasks are:
>
> - Hold periodic safety meetings with all employees.
> - Include safety status in all work group meetings.
> - Inspect department weekly and write safety work orders as required.
> - Have at least five one-to-one contacts regarding safety with employees each week.
> - Investigate injuries and accidents in accordance with Managing Safety guidelines within 24 hours.
>
> In addition, in agreement with Department Head:
>
> - Select at least two other tasks from a provided list and agree on what measurable performance is acceptable.
> - Report on these activities weekly.
>
> **WEEKLY SAFETY REPORT**
>
> The First Level Manager shall prepare and distribute a Weekly Safety Report in accordance with the format shown in Exhibit 4-6.
>
> **MEASURE OF PERFORMANCE**
>
> - Successful completion of tasks.
>
> **REWARD FOR PERFORMANCE**
>
> Safety will be listed as one of the key measures on the Accountability Appraisal Form.

Exhibit 4-4. First line supervisor accountabilities—Menu System.

A Menu System

Many organizations find SCRAPE too top-down in style and SBO too loose, and opt for a system halfway between the two, known as a Menu system. Here certain tasks are deemed to be mandatory and others optional, to be selected from a menu of activities provided by the organization. Exhibit 4-4 shows the Menu system for one organization; Exhibit 4-5 for another. Exhibit 4-6 shows the weekly report a supervisor might fill out to send his or her boss showing task completion.

These are examples at the first-line supervisory level. Obviously the accountability system must be in place at all other levels of the organization.

ACCOUNTABILITY SYSTEMS

ABC COMPANY
ACCOUNTABILITIES—FIRST LINE SUPERVISOR

- ACCOUNTABILITIES
  - Accident Investigation
  - Continuous Departmental Inspection
  - Employee Communications
  - Employee Training
- OPTIONAL ACTIVITIES
  - One-to-One Contacts
  - Group Meetings
  - STOP Program
  - Safety Improvement Teams
  - Positive Reinforcement
- PERFORMANCE MEASUREMENT
  - 100%, Based on Activities
- PERFORMANCE WEIGHTING
  - 20% of Total Performance Appraisal

Exhibit 4-5. First line supervisor accountabilities.

FIRST-LEVEL MANAGER'S WEEKLY SAFETY REPORT

From _____ To _____ Week Ending _____

1. Working Group Safety Meeting
   Date _____ Subject _____

2. Department Safety Inspections
   Date _____ Findings _____

3. One-To-One Contacts
   Employee _____ Date _____
   Employee _____ Date _____
   Employee _____ Date _____
   Employee _____ Date _____
   Employee _____ Date _____
   Employee _____ Date _____

4. Injury Status
   Name _____ Date _____
   Injury Description _____

5. Other Safety Tasks
   Description   Action   Compliance To Goal

Report Distribution: _____ Staff Manager
                     _____ Employee Relations Manager

Exhibit 4-6. First level manager's safety report.

> **ABC COMPANY**
> **DEPARTMENT MANAGER**
>
> - ACCOUNTABILITIES
>   - Assure Supervisory Performance by Receiving and Reacting to Reports
>   - Audit Performance Through Spot Checks
>   - Maintain Departmental Budget
>   - Visibly Participate in Programs
>   - Develop Safety Management Knowledge and Skills in Subordinates
> - OPTIONAL ACTIVITIES
>   - Participate in Audits
>   - Participate in Inspections
>   - Initiate One-to-One Contacts
>   - Create Ad Hoc Committees
>   - Support Recognition Programs
> - PERFORMANCE MEASUREMENT
>   - 25-50%, Numbers
>   - 50-75%, Audit & Activities
> - PERFORMANCE WEIGHTING
>   - 20% of Total Performance Appraisal

Exhibit 4-7. Department manager's accountabilities.

Middle Management

The basic performance model applies here also. Middle managers also must have clearly defined roles, valid measures of performance, and rewards contingent on performance sufficient to get their attention.

The performance at this level is critical to safety success. The middle managers (the persons to whom the first-line supervisors report) are much more a key than the supervisors, for they make the system run or allow it to fail. The middle manager's role is threefold:

- To ensure subordinate performance.
- To ensure the quality of that performance.
- To do some things that visually say safety is important.

Examples of accountability systems at the middle management level are shown in Exhibits 4-7 and 4-8.

Top Management

At the executive level we have a somewhat different ball game. While the basic performance model may be similar, it is certain that some inputs are more important at the top. The executive's individual traits (genuine executive interest) are a large factor, and role perception (believing he or she has a role) is crucial. Clearly, visibly demonstrated management commitment is essential. But what drives performance? The classic answer is

---

ABC COMPANY
DEPARTMENT MANAGER

GENERAL

The key accountability of the Department Managers is to ensure that the plans and programs of the XYZ Company Safety System are carried out in their area.

TASKS

- Review reports from their area on task accomplishments and act accordingly.
- Assess task performance defined for subordinate managers and feedback as appropriate.
- Engage in some self-defined tasks that can readily be seen by the work force as demonstrating a high priority to employee safety.
- Develop safety management knowledge and skills in subordinate managers.
- Make one-to-one safety contacts with hourly employees.
- Participate in department safety inspections.

PERIOD SAFETY REPORT

Department Managers shall prepare and distribute a periodic safety report in accordance with the format shown.

MEASURES OF PERFORMANCE

- Safety audit results for area(s) of accountability.
- The 13-period rolling total injury frequency record for area(s) of responsibility.

REWARD FOR PERFORMANCE

- Safety will be listed as one of the key measures on the Accountability Appraisal Form.

Exhibit 4-8. Department manager accountabilities and performance measures.

money—though that answer is probably less true than we think. Different executives are driven to safety performance for different reasons.

There is probably more interest and commitment at the executive level today than we have ever seen before. Perhaps the Bophal, Chernobyl, and Challenger incidents explain some of it. The safety professional who does not take advantage of this interest is missing a major opportunity.

While interested today, the executive typically does not have the foggiest idea of what to do to make safety happen. This is the safety professional's job—to spell out the role and to spell out the system.

The entire accountability system starts with a clear definition of roles.

ROLES

The roles of several layers of the line hierarchy are as follows:
1. **The role of the first-line supervisor is to carry out some agreed-upon tasks to an acceptable level of performance.**
2. **The roles of middle and upper management are to:**
   a. **Ensure subordinate performance.**
   b. **Ensure the quality of that performance.**
   c. **Personally engage in some agreed-upon tasks.**
3. **The role of the executive is to visibly demonstrate the priority of safety.**
4. **The role of the safety staff is to advise and assist each of the above.**

The supervisor's role as defined above is relatively singular and simple—to carry out the tasks agreed upon. What are those tasks? While it may depend upon the organization, the tasks might fall into these categories:

Traditional Tasks	Nontraditional Tasks
• Inspect	• Give positive strokes
• Hold meetings	• Ensure employee participation
• Perform one-on-ones	• Do worker safety analyses
• Investigate accidents	• Do force-field analyses
• Do job safety analyses	• Assess climate and priorities
• Make observations	• Perform crisis intervention
• Enforce rules	
• Keep records	

In addition to the above, supervisors no doubt will be responsible for certain day-to-day actions not easily spelled out or measured (following standard operation procedures [SOPs]). How well do they understand their responsibilities? They often simply do not know what is expected of them, particularly in safety. Almost always they have no idea of the extent of their authority, and usually they are unclear as to how their performance is being measured, again particularly in safety.

As mentioned earlier, the role of the supervisor is to engage regularly (daily) in some pre-defined tasks. What are those tasks? There are many that have been traditionally perceived as crucial and others thought today to be very meaningful while somewhat nontraditional. For the most part, all of this is based mostly on tradition and opinion, not on research. As the various tasks are discussed, we will attempt to describe the tasks and where possible will share whatever research there is on them.

What are the crucial tasks for a supervisor? What must a supervisor do regularly to achieve safety success? The answer to this is that we do not know. The research seems to say that there are no crucial "must-do" tasks. The research seems to say that it does not make any difference what a supervisor does, as long as something is done regularly, daily, to emphasize the importance of safety.

CHAPTER 5 The Supervisor's Job

WHAT IS THE SUPERVISORY JOB?

A LOT HAS BEEN WRITTEN about the job of the manager or supervisor. Much of it falls under a general heading of leadership and discusses what makes a good leader, what a good leader does, and so on. Our purpose here is not to discuss leadership, for the more we read about leadership the more confusing it gets. What we are interested in is what characteristics set off a good supervisor from a poor one.

Traits

What personality traits are needed for good supervision? We have no idea. There has been no direct research on this question, separate and apart from leadership generally. Many of the early books dealing with leadership included lists of the psychological characteristics essential to leadership. One list contained 19 traits; one, 31 traits; and another, 10 traits. The most frequently listed traits were intelligence, noted by ten studies; initiative, by six; extraversion and sense of humor, by five; and enthusiasm, fairness, sympathy and self-confidence, each by four.

When personnel managers discuss with other executives the kind of person who can fill a management vacancy, certain terms, irrespective of function or level, keep cropping up with monotonous regularity. Some of these terms are analytical, decisive, determined, acceptable. This indicates that most managers still accept the personality-trait approach to choosing good supervisors.

Today, most researchers into leadership have pretty well rejected the personality-trait approach. In general, results in this field would appear to support the notion that while certain personality characteristics are present in all leaders, these traits are also distributed among everyone else in the world.

How Leaders Are Chosen

Another factor to consider in looking at the traits of supervisors is how a person is actually promoted into supervision. During the 1950s Professor Melville Dalton, a sociologist at UCLA, made a fascinating analysis of managerial promotion in a firm with a total work force of about 8,000.[1] He analyzed the promotion scheme not in terms of formal company policies, but in terms of actual patterns of promotion. Extrapolating his findings still tells us a lot about what helps a person get ahead in business.

The company's official policy indicated that ability, honesty, cooperation, and industry were important criteria for advancement. In addition, such factors as age, employment background and service, formal education, and relevant training were often believed to be of importance.

Dalton, however, found that none of these factors were really pertinent. He found, instead, several *unofficial* requirements that were actually the key factors in managerial promotions: being a Mason, being of Anglo-Saxon or German ancestry, being a member of a local club, and being a Republican.

The ethnic requirement is particularly revealing considering that people of Anglo-Saxon or German origin made up only 38 percent of the surrounding community. Yet they controlled 85 percent of the company's advisory and directive forces. Dalton also studied management promotion in three other companies and found that similar unofficial requirements were at play.

There do not seem to be identifiable ideal traits for a supervisor. And even if there were, according to Dalton's work, it is doubtful if supervisors would actually be selected by these traits. All of this suggests that it isn't character or personality that makes an effective supervisor. Effort seems to be the major factor in supervisory success. If you know what to do and put forth enough effort, you can be a successful supervisor.

Key Elements of Supervision

One well-known management consultant, Edward C. Schleh, lists eight basic elements of supervision, whether it be at the foreman or the general manager level.

1. *Goal setting.* The first responsibility of any supervisor is to set goals for his people and to spell out exactly what he expects from them. Moreover, the definition of objectives should be in

terms of results, not just actions. This seems obvious but is very rarely done consistently.

2. *Training.* Second, a supervisor should be responsible for training his people. This is particularly important at the first-line supervisory level.

    There is a strong tendency for a foreman to refer the new worker to an experienced worker for training. This often results in poor training, because the older person is not held accountable for the instructing. The new person often develops bad work habits, which are hard to change later.

3. *Follow-up.* A third essential is the follow-up. While no supervisor should look over the shoulders of his people all the time, he must make sure things are going right. For example, a sales manager must follow up to make sure a salesman has understood and is practicing the right sales approach. An accounting department head must follow up to be sure that all the special requirements of posting are understood. A general manager must follow up to make sure that his foreman understood the changes in basic policy.

4. *Discipline.* Fourth, a supervisor must discipline. This task, of course, is usually thought of as a nasty one, and nobody really likes to exercise discipline, but it is and must be part of a supervisor's job at any level in the organization.

5. *Motivation.* A fifth duty is to stimulate your people. Everyone drags from time to time, and one of the functions of leadership is to help others regain their enthusiasm. A new supervisor sometimes thinks his job stops with delivering orders from above. If he operates in this way, he just does not get as much out of his people as he could.

6. *New Methods.* Another responsibility is the installation of new methods. Only the supervisor can get new methods into operation.

7. *Subordinate development.* Seventh, a company must rely on a supervisor to develop his people for future promotion. The foremen must develop their people; the superintendent must develop his foremen; the plant manager must develop his superintendents, and so on down the line. This is a basic responsibility of every managerial job.

8. *Fixing accountability.* Finally, it is the responsibility of any manager to call his people to account. Employees must be straightened out if necessary or complimented for an accomplishment. Just as the supervisor must hold his people

accountable for results and for certain procedures, he also is held accountable for results and for doing certain things. In later sections this will be discussed in detail, for the real key to the success of any safety program (or any other activity) is how well you perform, and this is dependent on how strictly your boss holds you accountable.

Your Responsibilities

As an introduction to what follows, we're going to ask you now to think about your job as a supervisor—what you do—what your tasks and responsibilities are in your company.

In the exercise at the end of the chapter, you will find a list of duties and responsibilities that are typical for some supervisors. In your company, on your job you may or may not be responsible for these—only you know. There are no right and wrong answers—the purpose of the exercise is to clarify your thinking.

In the next section we'll ask you to look again at the list to assess whether or not you have the authority to accomplish what you need to in each of your areas of responsibility. Then we'll ask you to look at the list a third time to determine whether or not you are held accountable for results in your areas of responsibility. If so, how are you being measured? From your answers to these questions, we can determine pretty well how important a task is to your company; for the tighter the measurement, the more important it is in management's eyes and the more effort you'll put into it.

When you have finished the exercise, take it to your boss and see how he views your responses. See what his opinion is on the "Don't knows." This discussion with him in itself should improve your supervision, for not only will you have a clearer idea of what your responsibilities are but you and your boss will be in agreement on what they are.

YOUR RESPONSIBILITIES AS A SUPERVISOR

Listed below are 48 tasks and responsibilities. (Some of them may not be a part of your job.) Put a check mark in the column under the head that best indicates your responsibility for the tasks listed at the left.

Usually you will have responses in all three columns. This is normal. Do not be dismayed to find you have a number of check marks in the third column. Everyone does.

	Yes	No	I Don't Know

New Employees
1. Hire .... _____ _____ _____
2. Accept or reject applicants .... _____ _____ _____
3. Report on probationary employees .... _____ _____ _____

Training Employees
4. Orient new employees .... _____ _____ _____
5. Explain safe operation/rules .... _____ _____ _____
6. Hold regular production meetings .... _____ _____ _____
7. Hold safety (tool box) meetings .... _____ _____ _____
8. Coach employees on the job .... _____ _____ _____

Production
9. Control quantity .... _____ _____ _____
10. Control quality .... _____ _____ _____
11. Stop a job in progress .... _____ _____ _____
12. Authorize changes in setup .... _____ _____ _____
13. Requisition supplies .... _____ _____ _____
14. Control scrap .... _____ _____ _____
15. Establish housekeeping standards .... _____ _____ _____

Safety
16. Take unsafe tools out of production .... _____ _____ _____
17. Investigate accidents .... _____ _____ _____
18. Establish inspection committees .... _____ _____ _____
19. Inspect your own department .... _____ _____ _____
20. Correct unsafe conditions .... _____ _____ _____
21. Correct unsafe acts .... _____ _____ _____
22. Send employees to the doctor .... _____ _____ _____

Discipline
23. Recommend promotions or demotions ..... _____ _____ _____
24. Transfer employees out of your department . _____ _____ _____
25. Change an employee to a less desirable job. _____ _____ _____
26. Grant pay raises .... _____ _____ _____
27. Issue warnings .... _____ _____ _____
28. Suspend .... _____ _____ _____
29. Discharge .... _____ _____ _____

Assigning Work
30. Prepare work schedules .... _____ _____ _____
31. Assign specific work .... _____ _____ _____
32. Delegate authority to leaders .... _____ _____ _____
33. Authorize overtime .... _____ _____ _____

Employee Affairs
34. Prepare vacation schedules .... _____ _____ _____
35. Grant leaves of absence .... _____ _____ _____
36. Lay off for lack of work .... _____ _____ _____

37. Process grievances . . . . _____   _____   _____

Coordination
38. Authorize maintenance and repairs . . . . _____   _____   _____
39. Make suggestions for improvement . . . . _____   _____   _____
40. Discuss problems with management . . . . _____   _____   _____
41. Recommend changes in policy . . . . _____   _____   _____
42. Improve work methods . . . . _____   _____   _____

Cost Control
43. Reduce waste . . . . _____   _____   _____
44. Keep production records . . . . _____   _____   _____
45. Budget . . . . _____   _____   _____
46. Approve expenditures . . . . _____   _____   _____

Other
47. Keep up employee morale . . . . _____   _____   _____
48. Reduce turnover . . . . _____   _____   _____

WHAT AUTHORITY DO YOU HAVE?

We hope the previous exercise generated some confusion in your mind about your responsibilities as a supervisor. And we hope your uncertainties were resolved with your boss so that you now have a clearer perception of your job and what you are responsible for. Shortly you'll go through a similar process to see if you have any confusions about your authority—that is, how much power do you have to get things done?

Formal Aspects of Authority

Let's start with some definitions, not to be academic but to try to clarify an important yet fuzzy concept. Sometimes when we talk of authority, we are thinking about something formal, something that is granted automatically by rank. Authority can be defined by one's military rank, for example. The captain may not know exactly how much authority he has or even what it is, but he knows he has more than the lieutenant and less than the major.

We can also define authority by calling it a kind of power. It is an institutional mechanism that aims to define which of two members in a relationship will be the superior. Authority is potential extra power, given by an organization to some of its members in order to guarantee an unequal distribution of power.

Sometimes the power thus delegated has nothing to do with relationships. For example, the organization may assign to someone the power to spend some of its money for supplies. But very often authority does include power over other people, power to restrict or punish, and power to reward.

Degrees of Authority

So far we have discussed authority as a black and white sort of thing—either we have it or we don't. But things are seldom black or white, and authority doesn't work that way. Seldom, for instance, do we have no authority in a situation. We can always extend some influence. On the other hand we rarely have complete authority over people or things. It is usually limited in terms of dollars we can spend, man-hours we can commit, and other factors.

To help you clarify how much authority you have on your job, we have defined five degrees of authority.

1. *Almost complete authority.* You can make any decision related to an area of responsibility and can implement that decision without having to check with anybody. You don't even have to inform anyone else.
2. *Decide and implement but keep them informed.* You can make a decision and you can implement your decision, but you must tell the boss what you've done.
3. *Decide, but check.* The decision is yours, but you cannot implement your decision until you've cleared it with the boss.
4. *Let the boss decide.* You can gather the facts but are required to have the boss make the decision.
5. *Influence only.* You have no authority except your personal influence.

These are not the only ways businesses classify power. Often companies define degrees of authority in more concrete terms, of dollars an individual can spend without approval. But the above five categories will work better for our purposes.

Advantages and Disadvantages in Using Authority

Industrial organizational structures seem to be designed with authority in mind. We build organizations in the shape of pyramids because that shape makes the exercise of authority easier. Pyramids create differences in rank and status, and the people in higher ranks can use their authority to influence lower ranks. Superiors in industrial organizations almost naturally turn to authority whenever a change problem arises with subordinates. The very idea of delegating authority rests on the assumption that authority can help people who have more of it to change the behavior of those who have less of it. In fact, we usually even define the superior in a relationship as the person with more authority.

From the manager's viewpoint, the advantages of authority are substantial. For one thing, it can be used immediately for an immediate result, without justification. You don't have to know much about any particular employee to be fairly certain that firing him or cutting his pay or demoting him will strike at some important needs and thereby keep him in line. But you might have to know a good deal about the same employee to find out how to make work more interesting for him, or to turn him on to work.

Another advantage is its directness. Authority does not require much subtlety or much understanding of people's motives. It is, for example, simpler to spank a child when he misbehaves than to distract him or to provide substitute satisfactions or to explain why what he did was wrong. Given a hundred children, it is easier to keep them in line by punishing a few than to teach them all to feel responsible for their own behavior.

Also, exerting authority is often personally gratifying to the superior and, therefore, attractive. Authority often fits neatly into our need to blow off steam. When you spank a child, you not only change his behavior but you also provide yourself with an outlet for the tensions which have built up within you as a result of problems and conflicts—with your boss, your wife, or anyone else.

Authority is sometimes seen as a way for a superior to guarantee his superiority. If his subordinates know that he can and will punish readily, they are likely to behave respectfully and submissively, at least in his presence.

Using authority has another advantage: speed. A do-it-now-or-else order usually gets it done, without having to waste time and energy on an explanation

Using formal authority also has some negative by-products. When a supervisor's activity interferes with an employee's efforts to satisfy his own important needs, the employee may not sit still very long for it. A manager often finds he has changed behavior he had not intended to change as well as (or instead of) behavior he did intend to change. The child who is spanked every time he puts his hand into the cookie jar may learn to keep his hand out, or he may learn to go to the jar only when you're not looking. He may also learn that you are out to keep him from getting what he wants, and what he thinks he needs.

Employees who are chewed out whenever they are caught loafing may learn to act busy when you're looking. You have become an enemy, and they are provided with a challenging game to play against you: Who can think up the best ways of loafing without getting caught. This is a game they feel justified in playing because of your actions. It also is a game they can invariably win. It is also fun to play.

Another difficulty with using your authority is that it may be irreversible. It is not as easy to pat a subordinate's head after spanking him as it is to spank him after patting, for human beings have memories. Using your authority tends to reduce feedback rather than build it. A series of bad experiences for an employee may well destroy the possibility of further communication and cooperation between him and you. And once you've lost him, you're in trouble. It will be extremely difficult to reestablish positive contact.

All in all, the advantages and disadvantages of using authority seem to suggest that restrictive methods may be effective only in situations that meet certain conditions. Such conditions might be (1) when you are trying to bring about a change in a specific action, rather than in generalized action or in attitude, (2) when restrictions are not seen by an employee as depriving or frustrating, and (3) when speed or safety or uniformity is critical to your task.

Keep in mind that we are talking about formal authority, the kind of power that has been officially delegated. Power to influence behavior also comes from other sources: your skills, your personality, your reputation. The exercise at the end of the chapter is designed to help you assess whether or not you have formal authority in certain situations. It looks at the same areas as those in the previous exercise and asks you to assess the degree of authority you believe you have in each area. When you have finished, take the exercise to your boss and discuss your responses.

YOUR AUTHORITY AS A SUPERVISOR

Listed below are the same tasks and responsibilities presented in the exercise earlier in this chapter. For each you may have some degree of authority to get things done, whether over people, over expenditures, or both. Put a check mark in the column under the head that best describes the degree of authority you have over the situations listed at the left.

Again you will usually have responses in all columns. Again you may have difficulty in deciding. If so, enter a question mark under the column which seems right.

	Complete Authority	Decide and Implement but Inform	Decide but Check	Boss Decides	Influence Only

New Employees
1. Hire.................................. ____ ____ ____ ____ ____
2. Accept or reject applicants............ ____ ____ ____ ____ ____
3. Report on probationary employees ..... ____ ____ ____ ____ ____

Training Employees
4. Orient new employees .... ____ ____ ____ ____ ____
5. Explain safe operation/rules........... ____ ____ ____ ____ ____
6. Hold regular production meetings ...... ____ ____ ____ ____ ____
7. Hold safety (tool box) meetings .... ____ ____ ____ ____ ____
8. Coach employees on the job .... ____ ____ ____ ____ ____

Production
9. Control quantity .... ____ ____ ____ ____ ____
10. Control quality .... ____ ____ ____ ____ ____
11. Stop a job in progress .... ____ ____ ____ ____ ____
12. Authorize changes in setup .... ____ ____ ____ ____ ____
13. Requisition supplies .... ____ ____ ____ ____ ____
14. Control scrap .... ____ ____ ____ ____ ____
15. Establish housekeeping standards...... ____ ____ ____ ____ ____

Safety
16. Take unsafe tools out of production ..... ____ ____ ____ ____ ____
17. Investigate accidents .... ____ ____ ____ ____ ____
18. Establish inspection committees .... ____ ____ ____ ____ ____
19. Inspect your own department.......... ____ ____ ____ ____ ____
20. Correct unsafe conditions............. ____ ____ ____ ____ ____
21. Correct unsafe acts .... ____ ____ ____ ____ ____
22. Send employees to the doctor .... ____ ____ ____ ____ ____

Discipline
23. Recommend promotions or demotions .. ____ ____ ____ ____ ____
24. Transfer employees out of your department .... ____ ____ ____ ____ ____
25. Change an employee to a less desirable job....................... ____ ____ ____ ____ ____
26. Grant pay raises.................... ____ ____ ____ ____ ____
27. Issue warnings..................... ____ ____ ____ ____ ____
28. Suspend........................... ____ ____ ____ ____ ____
29. Discharge......................... ____ ____ ____ ____ ____

Assigning Work
30. Prepare work schedules.............. ____ ____ ____ ____ ____
31. Assign specific work................. ____ ____ ____ ____ ____
32. Delegate authority to leaders .... ____ ____ ____ ____ ____
33. Authorize overtime .... ____ ____ ____ ____ ____

Employee Affairs
34. Prepare vacation schedules............ _____ _____ _____ _____ _____
35. Grant leaves of absence .... _____ _____ _____ _____ _____
36. Lay off for lack of work .... _____ _____ _____ _____ _____
37. Process grievances .... _____ _____ _____ _____ _____

Coordination
38. Authorize maintenance and repairs ..... _____ _____ _____ _____ _____
39. Make suggestions for improvement ..... _____ _____ _____ _____ _____
40. Discuss problems with management .... _____ _____ _____ _____ _____
41. Recommend changes in policy .... _____ _____ _____ _____ _____
42. Improve work methods............... _____ _____ _____ _____ _____

Cost Control
43. Reduce waste...................... _____ _____ _____ _____ _____
44. Keep production records .... _____ _____ _____ _____ _____
45. Budget .... _____ _____ _____ _____ _____
46. Approve expenditures .... _____ _____ _____ _____ _____

Other
47. Keep up employee morale .... _____ _____ _____ _____ _____
48. Reduce turnover.................... _____ _____ _____ _____ _____

HOW ARE YOU MEASURED?

We have one more process to go through which will round out, we hope, your picture of your job. This has to do with how you are held accountable for the various aspects of your job.

Let's define accountability as the fact of active measurement by management to ensure compliance with its will. In other words, accountability pins down the question of whether—and how—the boss is measuring you to see that you are carrying out the tasks he told you to do.

Types of Measurement

How does management measure your performance? Usually it's done with some kind of numbers game. In production, for instance, your department's performance is probably measured either by how many units it produces or by whether it meets a quota (or by a percentage of a quota). These are numerical measurements, used mostly in production, in quality control, and in cost control where it is easy to develop them.

However, it is not easy to use a numerical measurement for many areas of your job—in new employee orientation, training, discipline, or coordination for example. Therefore, management has to devise other measures, such as a spot-check method or a report from you that states whether or not you did the required task.

Tasks that are not easily measured with numbers are usually not measured as frequently or as well by management. This means that your boss has a much better handle on your production results than he does on your training effectiveness. Because of this, you probably spend considerably more time and effort on things that improve production than you do on things related to training. You know you can show results in production by concentrating on it. You can literally ignore training, safety, and other aspects for months, and management would never know the difference.

How you are measured dictates where you spend your time. The tighter the measure, the more you will pay attention to the area.

Results Versus Activity Measures

There are two other ways that management uses to measure your performance in a single area. It can measure the end results of your effort, or it can measure whether or not you are carrying out your defined tasks. An example is in safety. *Results measurement* asks how many accidents have happened to your people, how many days were lost by them, how much it cost the company. Activity measurement asks how many times you have talked to your people about safety or how many times you've inspected your department.

Activities measures tend to be tighter, more restrictive. Results measures tend to give you more leeway. They allow you to do whatever you want as long as you get results. Management often uses activities measures while you're learning a task. Once you are highly competent, it uses results measurements.

Degrees of Accountability

No one escapes being measured in some way on the job. How you are assessed depends on what your tasks are and how much authority you have. Just as there are degrees of authority, there are degrees of accountability. For instance, one company describes three levels of accountability.

1. *Strong accountability.* Must accept heavy contribution for the achievement and effectiveness of end results; requires active participation and the initiation of action.
2. *Medium accountability.* Must accept a secondary contribution to the achievement and effectiveness of end results; may be through an interrelated function or may be heavy in one facet of a many-faceted accountability. Also requires active

participation and the initiation of action within the area of the persons concerned.
3. *Light accountability.* Must accept accountability for contributing but not for initiating any action.

Your Job's Accountabilities

Look again on the next two pages at the 48 tasks and responsibilities in the previous two exercises. For each task you will be asked to determine: (1) If the boss is holding you accountable—actually regularly measuring your performance; (2) How he measures you in that area; (3) Whether or not the measurement is on your results or on your activities. The purpose again is to clarify your job.

If you are confused about any of the answers, meet with your boss and get the point clarified. It is crucial to your success to know how you are measured.

YOUR ACCOUNTABILITY AS A SUPERVISOR

For each task that is part of your job, identify whether or not your boss is, in fact, measuring your performance (yes or no). Briefly state how he measures it (numerical or other—state specifically the measure he uses), then indicate whether the measure is a results or an activities measure. Finally, check what degree of accountability is used: strong, medium, or light.

WHERE DOES SAFETY FIT INTO YOUR JOB?

We hope the three exercises helped you find out some things about your job that you didn't realize before. Supervisors going through this process usually do. But what does all this have to do with safety? Quite a bit, we think. First of all, one of the categories you looked at was safety. Under that category we listed seven specific tasks.

The Function of Safety

Before we look at your specific responsibilities and tasks in safety in the next section, lets look at the field of safety itself. What is safety and where does it fit into your job?

For many years the slogan of the National Safety Council was "safety first." Safety professionals believed and preached this for a long time. Today we realize that we really do not want "safety first" any more than we want "safety last." Either slogan means safety is being considered as something separate from the other aspects of production. Obviously, we want

	Measured		Type	Degree				
	Yes	No		R	A	S	M	L
New Employees								
1. Hire								
2. Accept or reject applicants								
3. Report on probationary employees								
Training Employees								
4. Orient new employees								
5. Explain safe operation/rules								
6. Hold regular production meetings								
7. Hold safety (tool box) meetings								
8. Coach employees on the job								
Production								
9. Control quantity								
10. Control quality								
11. Stop a job in progress								
12. Authorize changes in setup								
13. Requisition supplies								
14. Control scrap								
15. Establish housekeeping standards								
Safety								
16. Take unsafe tools out of production								
17. Investigate accidents								
18. Establish inspection committees								
19. Inspect your own department								
20. Correct unsafe conditions								
21. Correct unsafe acts								
22. Send employees to the doctor								
Discipline								
23. Recommend promotions or demotions								
24. Transfer employees out of your department								
25. Change an employee to a less desirable job								

YOUR ACCOUNTABILITY AS A SUPERVISOR

	Measured Yes / No	Type	Degree R / A / S / M / L
26. Grant pay raises			
27. Issue warnings			
28. Suspend			
29. Discharge			
Assigning Work			
30. Prepare work schedules			
31. Assign specific work			
32. Delegate authority to leaders			
33. Authorize overtime			
Employee Affairs			
34. Prepare vacation schedules			
35. Grant leaves of absence			
36. Lay off for lack of work			
37. Process grievances			
Coordination			
38. Authorize maintenance and repairs			
39. Make suggestions for improvement			
40. Discuss problems with management			
41. Recommend changes in policy			
42. Improve work methods			
Cost Control			
43. Reduce waste			
44. Keep production records			
45. Budget			
46. Approve expenditures			
Other			
47. Keep up employee morale			
48. Reduce turnover			

effective production first, but it should be accomplished in such a manner that no one is hurt and losses are minimized.

Previously, safety professionals were striving for "safety programs" for their companies. The aim was to superimpose a safety program on the organization. Today, safety professionals realize that what is really needed is "built-in" safety, or "integrated" safety, and not some artificially introduced program. Safety must be an integral part of a company's procedures. We do not want production and a safety program, or production and safety, or production with safety—but rather, we want safe production.

The goal of management in general is efficient production—production which maximizes profit. To obtain this goal it has two basic resources: (1) employees and (2) facilities, equipment, and materials. Management brings many influences to bear upon both of these resources: the company's personnel are affected by training, selection and placement processes, employee health programs, and employee relations; facilities, equipment, and materials are influenced by maintenance, research, and engineering. These influences and basic resources are brought together through various procedures.

The function of a safety program is to build safety into these procedures and continually audit the carrying out of these procedures to ensure that the controls are adequate. These tasks are accomplished by continually asking why certain acts and conditions are allowed and whether certain known controls exist. Exhibit 5-1 depicts the relationship of these forces graphically.

Safety is not just a resource, an influence, or a procedure, and it certainly is more than a "program." Safety is a state of mind, an atmosphere that must become an integral part of each and every procedure that the company has. This, then, is what we mean by "built-in," or "integrated," safety. It is the only brand of safety that is permanently effective.

Relating the concept of "integrated," or "built-in," safety to your job, merely means that you include safety as one of your many considerations in anything performed in your department. When you plan, give orders, train, follow up, or anything else, one of the aspects you consider is safety. This is a frame of mind more than it is another duty. Or stated better, if it is a frame of mind, it is not another duty for you.

Priorities

Many supervisors start out with the idea that safety is separate from production, rather than building safety into regular work systems and activities. This immediately places that supervisor in a "choice" situation. Daily, perhaps hourly, he must choose between safety and production. He then

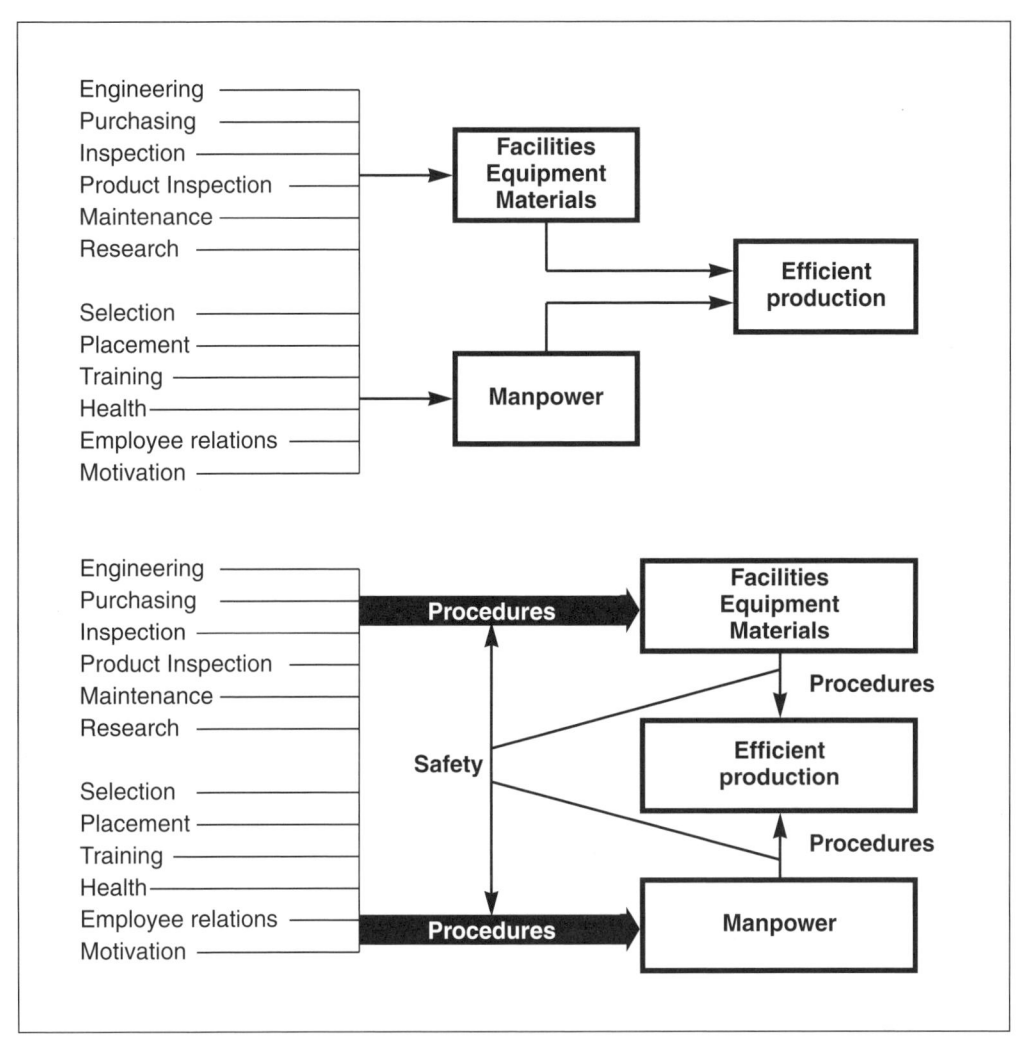

Exhibit 5-1. *Graphic representation of "integrated" safety. From* Techniques of Safety Management *by Dan Petersen. Copyright © 1987 by Dan Petersen.*

must give priority either to production or to safety. Obviously if a choice must be made, safety cannot win. Profit, defined as production, is the name of the game, the reason the manager exists; it is his job. We then have placed safety in a win-lose situation. By "building in" safety we do not force ourselves to make a choice—we strive for safe production, and we place ourselves in a win-win situation.

Notes

1. Melville Dalton, *Men Who Manage* (New York: Wiley, 1959).

CHAPTER 6 The Safety Job

HOW SAFETY GETS ACCOMPLISHED

A GOOD SAFETY RECORD DOES NOT JUST HAPPEN—it is the result of a number of people doing a number of things well. Taken as a whole these activities have been called a "safety program." Our industrial safety programs, then, are basically collections of recommended procedures and actions. These are typical components of safety programs:

Management's statement of policy	Motivational activities
Corporate safety rules	Inspections
Definitions of responsibility	Investigations
Screening of employees	Record keeping
Placement of employees	Record analyses
Training of employees (orientation)	First-aid training
Ongoing training	Medical facilities
Supervisory training	Others

Who Performs the Activities?

A number of people are involved in a "safety program."

Top management. Those who decide what they want done and then set direction or policy.

Middle management. The group in an organization that is located between the policymakers and the first-line supervisors.

Staff people. These are the various specialty people who have a function of assisting the top in setting policy, of working with the middle in often rather ill-defined ways, and of influencing the first-line supervisors in a number of ways to get them to want to do what the specialty people and management want done.

The supervisor. In the eyes of most safety professionals the first-line supervisor carries out the most critical functions. The second most critical functions are top management's.

A 1967 National Safety Council survey (and a 1992 re-survey) brought out a number of rather interesting points.[1] The purpose of the survey was to determine which factors are considered most important to a comprehensive industrial safety program.

One hundred forty-eight industries took part in the original survey by completing a questionnaire to rate the importance of various safety activities. A total of 78 activities in eight safety program areas were included. Industry people rated the importance of the eight major areas, as well as the groups of activities within each area. The major program areas and the top-rated activities in each are as follows:

1. *Supervisory Participation* (SP)
   Enforcing safe job procedures.
   Setting an example by safe behavior
   Training new or transferred employees in safe job procedures.
2. *Middle Management Participation* (MP)
   Setting an example by behavior in accord with safety regulations.
   Restating management's position on safety.
   Using safety as a measure of management capability.
3. *Top Management Participation* (TP)
   Setting an example by behavior in accordance with safety regulations.
   Assigning someone to coordinate safety on a full- or part-time basis.
   Publishing a policy expressing management's attitude on safety.
4. *Engineering, Inspection, Maintenance* (EIM)
   Specifying guards on machinery before it is purchased.
   Setting up a formal lockout procedure.
   Establishing a system of preventive maintenance for tools, machinery, plant, and other items and areas.
   Inspecting tools and equipment periodically.
5. *Screening and Training of Employees* (ST)
   Making safety a part of every new employee's orientation.
   Including safety in supervisory training courses.
   Including safety requirements in job procedures based on job safety analyses.
6. *Coordination by Safety Personnel* (CSP)
   Advising management in the formulation of safety policy.
   Analyzing the safety program to determine its effectiveness.
   Assisting and advising other departments on various safety-related matters.
7. *Forming a Record Keeping System* (R)
   Requiring the department supervisor to conduct investigations of disabling injuries.
   Using a standardized injury investigation form.
   Including recommendations in injury statistics reports.

8. *Motivational and Educational Techniques* (ME)
   Providing employees a list of general safety rules.
   Establishing a procedure for disciplining violators of safety rules.
   Holding work-place safety meetings.

The rank order of the major safety program areas is as follows:

- Supervisory Participation
- Top Management Participation
- Engineering, Inspection, Maintenance
- Middle Management Participation
- Screening and Training of Employees
- Records
- Coordination by Safety Personnel
- Motivational and Educational Techniques

The first five areas represent what might be considered the basics of a safety program, while the rest are somewhat peripheral. The ten sub-items from the major program areas that the survey participants felt were most important are:

- Enforcing safe job procedures. (SP)
- Setting an example by safe behavior. (SP)
- Middle management setting an example by behavior in accord with safety requirements. (MP)
- Training new or transferred employees in safe job procedures. (SP)
- Making safety a part of every new employee's orientation. (ST)
- Top management setting an example by behavior in accordance with safety regulations. (TP)
- Top management assigning someone to coordinate safety on a full or part-time basis. (TP)
- Including safety in supervisory training courses. (SP)
- Top management publishing a policy expressing management's attitude on safety. (TP)
- Advising management in the formulation of safety policy. (CSP)

The emphasis in the survey results is on supervisory and top management participation. This indicates that most people saw the supervisor as the crucial link directly affecting employee behavior. Top management must provide the initial push of a safety program, the supervisor must maintain program momentum daily, and middle management must participate to create the chain of communication and command.

Top management
Other than the first-line supervisor, the man considered most important to any safety effort is the big boss. His role in the safety program is to:
— Issue and sign safety policy.
— Receive information regularly as to who is and who is not doing what is required in safety as determined by some set criteria of performance.
— Initiate positive and/or negative rewards for that performance.

Systems for getting information to the boss regularly usually comes from the supervisor's regular reports, as discussed in the next chapter.

Middle management
The industrial safety survey also showed the importance of middle management participation. The list giving the rank order of the major safety program areas indicates that the role of the mid-manager is more important to safety success than the screening or training of employees, the record-keeping function, the coordination by the safety manager, and all motivational and educational techniques. In the survey the third most important item on the list of 78 was "middle management setting an example by behavior in accord with safety requirements." The survey was completed in 1967. The survey was replicated in 1992. The results were almost identical. The main difference was that employee involvement and participation had become the fourth most important category, following top management visibility, supervisory action, and middle management participation.

The survey also spelled out what the role of the mid-manager might be in safety: it is his role to restate policy, to participate in safety meetings, to review employee safety performance, to establish checks to ensure adherence to safety program goals, to set an example, to utilize safety performance as a measure of management capability, and to serve on investigating committees. In short, he is to be an active participant in the program by transforming the executive's abstract policy into supervisory action.

Safety manager
What a safety manager does is crucial to the success of a safety program in any company. First of all he should structure systems of measurement so that accountability can be fixed and rewards can be properly applied to the right people at the right time to reinforce the desired behavior. Some of these systems might incorporate estimated costing, sampling, rating effort, performance measurements, and others described later.

Secondly, the safety manager must be a programmer: a person who oversees those aspects of the safety program that are not completely under his control. He must make sure that safety is included in orientation, that safety training is provided where needed, that safety is a part of supervisory development, that things are done to help keep the organization's attention on safety, and that safety is included in employee selection, in the medical program, and in other areas.

Thirdly, he must be a technical resource, knowing how to investigate in depth where to get technical data, what the standards are, and how to analyze new products, equipment, and problems.

Fourthly, he must fill the role of systems analyst, searching for reasons why accidents happen and whether or not proper controls are in effect.

The supervisor

The major role of the safety program belongs to the first-line supervisor. Everything that everybody else does is worthless if the supervisor does not do his job. What is his job in the safety program?

While this may vary from company to company, there are four key tasks that belong to the supervisor in every safety program.

1. *Investigating* all accidents to determine underlying causes.
2. *Inspecting* his area routinely and regularly to uncover hazards.
3. *Coaching* (training) his people so they know how to work safely.
4. *Motivating* his people so they want to work safely.

Each of these tasks will be covered in detail in following chapters because these activities are fundamental to every safety program.

The traditional picture of a manager or supervisor shows him as having a nicely bounded, carefully defined, and compartmentalized role. This approach is characterized by organization charts and by such terms as "delegation," "line," and "staff." However, studies show clearly that successful managers in modern organizations operate in a manner very different from that suggested by the traditional view.

Leonard Sayles, in *Managerial Behavior*,[2] discusses some of the "old wives' tales" of management theory and shows how each is not valid in actual business practice.

A manager should take orders from only one man, his boss. Most managers, in fact, work for or respond to many people (customers or those in a position to make demands upon them).

The manager does no work himself; he gets things done only through the activities of his subordinates. Actually the manager himself must carry on many relationships with others and participates in activities of all kinds to get things done.

The manager devotes most of his time and energy to supervising his subordinates. In reality, the manager is away from his subordinates a significant portion of the time.

The good manager manages by looking at results. Normally, methods of continuous feedback are required.

To be effective, the manager must have authority equal to his responsibility. Actually a manager almost never has authority equal to his responsibility. He must depend on the actions of many people over whom he has not the slightest control.

Staff people have no real authority since they are subsidiary to the line organization. Actually staff groups have very real power.

The facts of organizational behavior make it perfectly plain that the modern manager operates in a continually changing context where things are not as simple as the "principles of management" might suggest.

In the last thirty years, these thoughts have become particularly obvious. There are fewer managers than ever before in organizations due to corporate "rightsizing." There are fewer supervisors—the span of control is invariably greater. Supervisors are relabeled "team workers" or "facilitators." There is less one-on-one contact with employees, as there "just isn't time any more." There is more paperwork (CYA stuff) or computer work. The Deming concept asks for a concentration on upstream measures to ensure "process improvement," etc.

All of this changes the role of the supervisor—may even shift that role to work teams, and will be discussed later in this chapter.

THE SUPERVISOR'S ROLE IN SAFETY

A fundamental principle in business has always been that the first-line supervisor is the key man in any safety program. Whether or not you accept this key role, however, is not always a sure thing. It depends on many factors.

The attitude of the majority of supervisors today lies somewhere between total acceptance and flat rejection of safety and their role in the safety program. Most typical is the organization in which line managers do not shrink from this responsibility, but at the same time do not really accept it either. Nor do they treat it as they would any of their other defined production duties and tasks. Usually their "safety hat" is worn less frequently than their "cost control hat," or their "production hat," or their "quality control hat." In most organizations safety is not considered to be as important to you, the line manager, as many, if not most, of the other duties that you are required to perform.

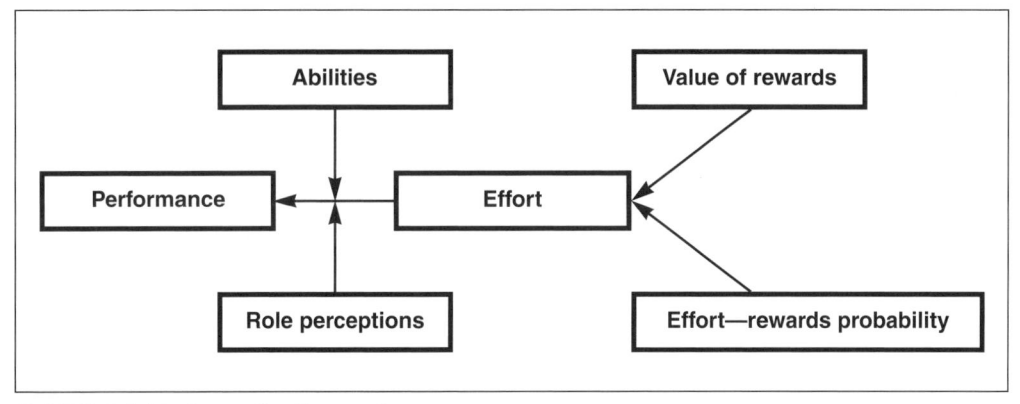

Exhibit 6-1. Factors affecting performance.

Your attitude toward safety and your performance in safety depends on three things: (1) whether or not you are able to perform, (2) whether or not you believe performing in safety is a part of your supervisory job, and (3) whether or not you try. (See Exhibit 6-1.)

In most cases today it is well worth the time and effort of the supervisor to concentrate on safety. You are key in the corporate safety program and management knows it. Management, because of this knowledge, will be measuring your safety performance in well-defined areas, which will undoubtedly include at least these:

— Whether or not all accidents are investigated and investigated to the point that root causes are found and removed.
— Whether or not routine, regular inspections are made to locate and remove hazards.
— Whether or not new employees as well as experienced ones are regularly and routinely coached to improve their performance.
— Whether or not all employees are motivated to want to work safely.

SAFETY WHERE THERE IS NO SUPERVISOR

During the last ten or fifteen years, management has undergone some amazing transformations. The changes in management thinking have been fast and furious in recent years; as each new book is published, management changes its direction and philosophy on how to manage.

The transition in management thinking started a number of years ago when managers began to question the dictates of "Classical Management," or "Scientific Management," as described in the early years by Frederick W. Taylor.

Following the classical school of management thinking came the "Human Relations" school of thought. To a degree, the human relations approach, popular in management until the 1970s and still taught today, was an outgrowth of the misinterpretations of the research of the 1920s, '30s, '40s, and '50s. Many interpreted the research to say "happy" workers are productive workers, thus the role of management was to make workers happy. Actually the research did not say it quite that way, as it is considerably more complicated than that.

The underlying assumptions of the above two schools of management thought were similar, but slightly different. Classical Management is based on the assumption that "everybody is alike—we can get the behaviors we want through manipulation." Thus we used wage incentive plans, piece rates, and other schemes to get more productivity. Human Relations Management has an underlying assumption which says, "Everybody is alike—we can get the behaviors we want by making them happy." Neither assumption had any basis in fact.

In the 1970s the "Contingency School of Management Thinking" emerged, and for the first time, management's underlying assumptions about people were in tune with psychological reality. The assumption behind the Contingency Theory of Management is that "everybody is different," therefore our management style, and how we deal with a worker, must be contingent upon the situation, upon the workers, and upon their needs. How a manager manages must be appropriate to the situation.

The "Situational Leadership" approach of Hersey and Blanchard became very popular in the late 1970s and early 1980s when one of the original authors, Blanchard, restated the basic thesis in different terms in the book, *The One Minute Manager*, which became a best seller.

The "Cultural Assessment" approach to management became popular in the '80s with the publishing of *In Search of Excellence*. This became one of the most popular management books in years. The authors quite simply looked at the companies in this country that consistently seem to be the best performers, the most effective in terms of bottom line performance. From this in-depth look at the best run companies, they distilled some keys to effective management.

Up to this point in our management philosophy evolution, we could remain fairly comfortable, for we could still try to feed our Classical Management style safety programs into our organizations. Then came the 1990s and new and different trends in management became popular. Dr. Deming's philosophy and the Total Quality Management concepts began to threaten our comfort levels, for now we were talking employee empowerment, no rule setting by management; using Statistical Process Control tools that could literally make much of our old safety tools obsolete.

In addition, "right sizing" and "re-engineering" in some companies had the effect of putting more and more work on fewer and fewer supervisors, managers, and employees.

In this process of attempting to improve corporate performance, one of the emerging trends has been to decide that there are levels of the traditional management hierarchy that we can do without. In some organizations this has meant the flattening of the organizations—reducing the number of middle managers and the number of layers of the hierarchy. This approach makes great sense in the age of computerization and engineering. In flattening the structure, we have become more efficient, more productive.

In other organizations, a decision has been made to go to team management, eliminating first line supervisors. This decision has great historical underpinnings—we have known since the old Hawthorne studies of the 1920s that the most productive work units are those where there is no formal supervision. It might be remembered, however, that in these early studies, team leadership was only an experiment—and as a result, people responded very positively (the Hawthorne Effect as it was later called).

As we go from supervisor control to team control, how do we ensure an accident/illness-free operation? The team approach requires us to reassess and reunderstand how safety can be accomplished in any organization.

The Total Involvement Concept

First, the "High-Performance Team Concept" needs some definition. It is different from what we have been talking about in recent years when we talk worker participation, involvement, or empowerment. The team concept suggests that decisions will be made and problems will be solved by work teams, not by individuals—as in empowerment. Empowerment, participation, and involvement can all happen in a hierarchical structure, when management simply asks the workers to help them in managing the organization, or one function of it, such as safety.

In a traditional, but worker empowered organization, little is changed when it comes to safety. We still have clear accountability within the management structure for proactive, daily, performance of those things that prevent incidents, although workers may do many of them. But accountability remains clear—it remains in line management from CEO to first-line supervisor.

The Team Concept

In the team concept, however, things have changed. First, there is no line supervisor to hold accountable through the traditional management tool of measurement coupled with rewards. When it comes to safety, there are other things to be considered.

Let us consider a few of our key safety principles in a team environment:
- The foreman is the key man (Heinrich, 1931). Today, there may not be a foreman.
- The Three E's of Safety (adapted from Heinrich, 1931)—Engineering, Education, and Enforcement. Two of these were traditionally the foreman's responsibility.
- Safety should be managed. (Petersen, 1970)
- The key is accountability (Petersen, 1970). How do we measure and reward a team?
- Management starts with a clear definition of roles. What is the role of the team?
- Who will do the traditional supervisory tasks of inspecting, investigating, motivation, training, one-on-ones, behavior observations, etc.?

OUR ATTEMPTS TO DATE

As we jumped into high performance approaches, we seem to have opted for one of several approaches, based on these assumptions:

1. We assumed that the teams will automatically take care of safety along with everything else.
2. In many cases we also assumed that by merely removing a person (the supervisor) that teams were ready for self (team) management.
3. We assumed that we in management should treat the team as we would a supervisor, saying "You are responsible for your own safety, now do something about it."
4. We assumed with employees now in charge of safety, that management was "off the hook," and no longer had a role—they abdicated.
5. We assume all people were ready to "take over." Some weren't. Some are happy to just do their thing.
6. We assumed all employees wanted a piece of the action. Some didn't. We created a three-tiered organization: (1) management; (2) involved workers; and (3) uninvolved workers, creating in some places another friction point.

7. We sometimes assumed if it didn't work, we could always go back and take back the authority. However, when we did this, it resulted in considerable chaos (a mild description). Occasionally we found it worked brilliantly, resulting in unheard of success.

We've identified some of the pitfalls; now, some of the ideas on how to achieve success with team approaches.

SUCCESSFUL TEAMS

Many organizations have been highly successful in utilizing the team approach. I've seen small units who have made the transformation with remarkable results. In talking with team members in successful companies in a variety of industries (food processing, oil drilling, railroads, steel companies), I have never seen such excitement from hourly workers as those experiencing team participation. And, without fail, the resultant safety records have been superb.

I've also seen the opposite—organizations where going to team approaches has produced problems, discontent, reduced morale, increased friction, and occasionally utter chaos.

What are the differences?—probably a number of things: readiness for participation; the planning that went into the change; the amount of involvement the workers had in the change decision and process, and other factors.

In those organizations that achieved the desired results, here are some things that managements have done—things that your organization might well consider:

1. Define within management exactly where you are willing to go by defining what you mean by "participation." Is it to get better input so management can make better decisions? Is it to share decision making? Is it to turn decision making over to the teams? And which decisions will management retain?
2. As best you can, check the maturity level of everyone who will be involved both in management and in production. Find out if you are "ready." This can be done through interviews or perception surveys. Also, look at the degree you've allowed self-management in the past before going to teams.
3. Check the amount of confidence and trust that exists between management and the workers. Culture surveys can help in the assessment.

4. Clearly define the ground rules, i.e., the guidelines defining what is allowable in decision making and what is not (regulatory compliance, core corporate values, etc.).
5. Allow great flexibility within those guidelines. For instance, "You must have a system on your team for recording and evaluating behavior observations. How you do it is up to you."
6. Require each location/team to have an annual (or periodic) review of their safety plan. Have an annual (or periodic) check of each unit to see whether or not they are carrying out the plan.
7. Hold each location/team accountable for what they have agreed to do.
8. Determine what core elements should be dealt with in each plan, but allow great flexibility in what each location/team can do to achieve the expected results within those core elements.
9. Require intermediate self-checks in shorter intervals; making frequent adjustments as needed.

AN EXAMPLE[3]

The Proctor & Gamble Company has been involved with team approaches and high employee involvement for many years and has experienced excellent results. While turning over much of the safety efforts to employee teams, they have set the ground rules for locations/units/teams by identifying "Key Elements" that teams must deal with.

Gene Earnest, former Senior Manager, Corporate Safety for Proctor & Gamble explained:

> About twelve years ago, the P&G corporate safety group developed what is known as the Key Elements of Industrial Hygiene & Safety. In effect, these are the "what counts" activities for IH&S and are the basis for site surveys. It was believed that if these activities were effectively implemented, injuries and illness would be reduced, and conversely, if they were done poorly, injuries and illness would increase. Because line management was involved in the development of this list, there was "buy-in."
>
> I cannot stress enough the importance of having a clearly identified IH&S program against which goals can be established at all levels of the organization and people held accountable for before-the-fact measures of injury and illness prevention. I recognize that certain areas of safety are "soft" and do not readily lend themselves to measurement; however, injury and illness producing events will

Element	Profile Rating	Priority
I. ORGANIZATIONAL PLANNING AND SUPPORT		
A. EXPECTATIONS AND INVOLVEMENT		
B. GOAL SETTING/ACTION PLANNING		
II. STANDARDS AND PRACTICES		
A. STANDARD IMPLEMENTATION		
B. SAFE PRACTICES		
C. PLANNING FOR SAFE CONDITIONS		
III. TRAINING		
A. SITE TRAINING SYSTEMS		
IV. ACCOUNTABILITY AND PERFORMANCE FEEDBACK		
A. BEHAVIOR OBSERVATION SYSTEM/SAFETY SAMPLING		
B. BEHAVIOR FEEDBACK		
C. PERFORMANCE TRACKING		

Exhibit 6-2. Industrial hygiene and safety key elements.

invariably be linked to behavior (of management and/or non-management) that is observable and measurable. What is being measured will vary depending upon the role of the individual.

The Key Elements of IH&S referred to previously divided into four major categories and nine subcategories (as shown in Exhibit 6-2).

Each key element is rated by the surveyor utilizing a scale of zero to ten where 0 means "nothing has been done" and 10 means the key element is "fully implemented and effective." Calibration to ensure that assigned ratings on surveys do not vary plus or minus one point is accomplished by corporate liaison accompanying the business sector surveyor on select surveys worldwide. Training and qualification programs for site health and safety resources also include exercises to ensure that people entering the role fully understand the industrial hygiene and safety key elements and how the ratings are assigned and calibrated.

The validity of this approach is evident from the graph (Exhibit 6-3) that illustrates the correlation between key element ratings and total incidence rates.

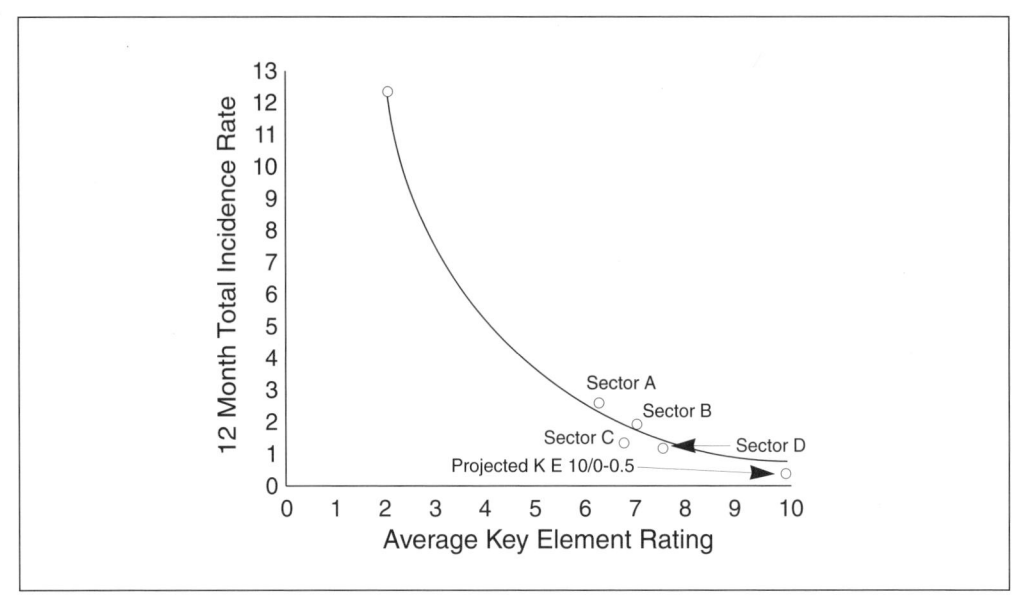

Exhibit 6-3. Acquisitions: Key element ratings vs. total incident rate.

New acquisitions have typically had total incidence rates (TIRs) of 12 and above, and Key Element ratings on the initial survey of 2-3. Established business sectors (A through D) have considerably higher Key Element ratings and correspondingly lower TIRs. The process of phasing in acquisitions typically requires several years. The projected TIR for a facility having a Key Element rating of 10 is 0-0.5.

Teams may deal with the areas in ways appropriate to, and compatible with, the culture of the team, but they shall be dealt with.

Then some criteria have been established for each. For instance, under the Key Element "Expectations and Involvement," a number of criteria are provided to measure the teams' accomplishments against, such as:

1. Health and safety responsibilities and expectations are established for roles/individuals within each level of the organization.
2. A performance feedback system exists and is utilized to hold each person accountable for meeting his/her responsibilities/expectations.
3. Job performance in the area of health & safety is perceived to have an effect (good or bad) on a person's career (pay, selection, promotion) the same as cost, quality, etc.

Under key element "Behavior Observation System/Safety Sampling," a number of criteria are provided, such as:

1. **Program structure**
    — Each unit has a formal Behavior Observation System (BOS) with a "system owner" identified.
    — All employees have been trained in the purpose, structure and management of the BOS.
    — All employees at all levels of the organization make observations.
    — Observations focus on behaviors. Behaviors listed on observation form are focused to:
        ‣ Monitor critical behaviors and this is periodically updated.
        ‣ Reduce target injuries/illnesses or monitor potential injury/illness sources.
        ‣ Monitor new policies and procedures.
    — Behaviors listed are clearly safe/unsafe (no gray areas)
    — Each person makes observations on a regular, frequent, and planned schedule at random times.
    — People are trained in the observation cycle and observation techniques.
    — Observations are made at random times; at least one BOS survey/shift (recommended).
    — Percent safe behavior correlates with injury and illness results.

After unit expectations are clearly established, it is up to the unit to perform. New acquisitions are given a defined period to come up to established expectations in the Key Element areas.

Units are measured in several ways: Total Company Incidence Rate is one way. Another is how well the unit is achieving progress in the Key Elements. A Key Element Rating is assessed on each unit periodically (see Exhibit 6-4). Goals are set in both areas. A corporate-wide goal was to achieve an 8+ KER/1.5 TIR at all sites.

As a result of concentrating on Key Elements, and requiring them at each location, the corporation has continued its decline in the TIR, recently reaching the lowest in history.

TE/DEPARTMENT _____ **DATE** _____

	Rating	Priority
I. ORGANIZATIONAL PLANNING AND SUPPORT		
A. EXPECTATIONS AND INVOLVEMENT		
B. GOAL SETTING/ACTION PLANNING		
II. STANDARDS AND PRACTICES		
A. STANDARD IMPLEMENTATION		

• Handling Chemicals: Key Ingredients, Hazardous Sys., Chemical Mgmt.
      • Life Threatening/Major Business Interruptions
      • Potentially Serious		

B. SAFE PRACTICES		
C. PLANNING FOR SAFE CONDITIONS		
III. TRAINING		
A. SITE TRAINING SYSTEMS		
IV. ACCOUNTABILITY AND PERFORMANCE FEEDBACK		
A. BEHAVIOR OBSERVATION SYSTEM/SAFETY SAMPLING		
B. BEHAVIOR FEEDBACK		
C. PERFORMANCE TRACKING		
OVERALL RATING		

OVERALL RATING

0   2   4   6   7   8   9   10

Key Element Ratings

0• Nothing has been done.
2• Some attempt has been made but no effective implementation.
- Might have had some effort/focus in the past but activity is not evident.
- Major system deficiencies that need to be addressed.

4• Systems are unsatisfactory/unreliable, much room for improvement.
- A system may exist and results tracked but it is not widely understood or communicated.
- No system owners exist, results are due to individual effort.

6• Partially implemented with several basics in place.
- System owner identified and is held accountable.
- Systems(s) beginning to deliver results but not sustained over time—first 6 months of system implementation.
- A correlation of results with action plans is not established.

8• Implemented and effective.
- Systems delivering solid results and have been sustained longer than 6 months.
- Intent of Key Element area(s) being met and there are no major outages.

10• Superior sustained long term results with full system and results documentation.
- It would be difficult to identify systems(s) improvements.
- Model system(s)/point of excellence approaching perfection.

Exhibit 6-4. Industrial hygiene and safety key elements.

Notes

1. All information about the survey results is from "Industrial Safety Study," an article by T. Planik, G. Driesen, and F. Valardo in the August 1967 issue of *National Safety News*. Copyright © National Safety Council. Used with permission.
2. Leonard Sayles, *Managerial Behavior* (New York: McGraw-Hill, 1964).
3. Excerpted from an article by R. Eugene Earnest, "What Counts in Safety," *Insights Into Management* 6 (Second Quarter, 1994): 2-6.

CHAPTER 7 Key Safety Tasks

INVESTIGATING FOR CAUSES

IN THIS PART we will be talking about four specific tasks that you as a supervisor must perform if a safety program is to work. They are:
- Investigating accidents to determine causes.
- Inspecting your area to identify hazards.
- Coaching your people to perform better.
- Motivating your people to want to work safely.

These are core tasks—musts—for a safety program to be effective. There are others that you should carry out and perform effectively. These will be discussed in a later part. These must be considered as "in addition" to the above four core tasks.

Investigating

Accident investigation is a device for preventing additional accidents.
According to the National Safety Council's *Accident Prevention Manual for Business and Industry*, 11th Edition, the principal purposes of an accident investigation are:
- To determine direct causes
- To uncover contributing accident causes
- To prevent similar accidents
- To document facts
- To provide information on costs
- To promote safety

The form shown in Exhibit 7-1 for the supervisor's investigation concentrates on the identification of the unsafe mechanical/physical/environmental condition and the unsafe act. The identification of these is based on a theory of accident causation that is often stated like this:

SUPERVISOR'S ACCIDENT REPORT   OSHA File _____
(To be completed immediately after accident, even when there is no injury)   (Yr. – Mo.)

Company name and address _____

Business SIC code _____

Plant or location address _____
(if different from above)

Accident date _____/_____/_____   Time _____

Day of week   ☐ Sun.   ☐ Mon.   ☐ Tue.   ☐ Wed.   ☐ Thur.   ☐ Fri.   ☐ Sat.

1. Name and address of injured (or ill) person _____
   _____
   SSN_____ 2. Age _____ 3. Sex _____

4. Years of service _____ 5. Time on present job _____ 6. Title/occupation _____

7. Department _____ 8. Location _____

9. Accident category (check) ☐ Motor vehicle ☐ Property damage ☐ Fire ☐ Other _____

10. OSHA recordable? ☐ Yes ☐ No

11. Severity of injury ☐ FATALITY ☐ Lost workdays ☐ Restricted workdays ☐ Medical treatment
    ☐ First-aid, return to work ☐ Other, specify_____

12. Amount of damage $ _____

13. Estimated number of days away from job _____

14. Nature of injury or illness _____

15. Part of body affected _____

16. Degree of disability ☐ Temporary total ☐ Permanent partial ☐ Permanent total

17. Describe in detail what happened _____
    _____
    _____
    _____
    _____
    Witness(es) _____

18. Describe any unsafe mechanical/physical/environmental condition at time of accident (be specific)
    _____
    _____
    _____

19. Describe any unsafe act by employees or others. (Be specific; must be answered.)
    _____
    _____
    _____

Exhibit 7-1. Sample supervisor's accident report form. (Page 1 of 2.)

20. Describe any personal factors that contributed to accident.

_____
_____
_____
_____

21. What applicable personal protective equipment (PPE) was being used? _____
_____ Did PPE fail? ☐ Yes ☐ No

22. What personal protective equipment should have been used? _____
_____ Was proper PPE available? ☐ Yes ☐ No

23. Corrective actions: what can be done to prevent a recurrence of similar incidents? _____

_____
_____
_____
_____

24. Describe in detail what happened. (Specify all machinery, chemicals or tools involved.) _____

_____
_____
_____
_____
_____
_____
_____
_____
_____
_____
_____
_____
_____
_____
_____
_____

PREPARED BY	DATE: / /	APPROVED BY	DATE: / /
NAME _____		NAME: _____	
TITLE _____		TITLE _____	
SIGNATURE _____		SIGNATURE _____	

Exhibit 7-1. Sample supervisor's accident report form. (Page 2 of 2.)

The occurrence of an injury invariably results from a completed sequence of factors, the last one of these being the injury itself. The accident that caused the injury is in turn invariably caused or permitted directly by the unsafe act of a person and/or a mechanical or physical hazard.[1]

This is known as the Domino Theory of Accident Causation. It urges us to find an act and/or a condition behind every accident and remove it. This approach is quite limiting, however. For today, we know that behind every accident there lie many contributing factors, causes, and subcauses. This newer theory, called Multiple Causation, states that factors combine in random fashion to cause accidents.

Let us briefly look at the contrast between the Multiple Causation Theory and the Domino Theory. Take for example this common accident: a man falls off a stepladder. If we investigate this accident using the Domino Theory, we are asked to identify one act and/or one condition:

- The unsafe act: climbing a defective ladder.
- The unsafe condition: a defective ladder.
- The correction: getting rid of the defective ladder

Investigating the same accident in terms of Multiple Causation, we ask about some of the contributing factors surrounding this incident.

- Why was the defective ladder not found in normal inspections?
- Why did the supervisor allow its use?
- Did the injured employee know he should not use it?
- Was he properly trained?
- Was he reminded?
- Did the supervisor examine the job first?

The answers to these and other questions would lead to the following kinds of corrections:

- An improved inspection procedure
- Improved training
- A better definition of responsibilities
- Pre-job planning by supervisors

With this accident, as with any accident, we must find some fundamental root causes and remove them if we hope to prevent a recurrence. Defining an unsafe act of "climbing a defective ladder" and an unsafe condition of "defective ladder" has not led us very far toward any meaningful safety accomplishments. When we look at the act and the condition, we look at symptoms, not at causes. If we deal only at the symptomatic level,

Name of Injured _____ Date of Accident _____ Time _____

Seriousness: ☐ Lost Time   ☐ Doctor   ☐ First Aid Only   ☐ Near Miss

Nature of Injury _____

What Happened? _____

What acts and conditions were involved *(use back also)*?
What caused them? How were they corrected?

	Unsafe Act/Condition/ Symptom	Possible/Probable Cause	Correction/ Suggested Correction
1.			
2.			
3.			
4.			
5.			

Supervisor _____ Department _____

Exhibit 7-2. Supervisor's report of accident investigation.

we end up removing symptoms but allowing root causes to remain to cause another accident or some other type of operational error.

Root causes often relate to the management system. They may be due to management's policies and procedures, supervision and its effectiveness, or training. Root causes are those that would effect permanent results when corrected. They are those weaknesses that not only affect the single accident being investigated, but also might affect future accidents and operational problems.

In this section we'll ask you to investigate accidents under the Multiple Causation Theory rather than the Domino Theory. The form used should be similar to the one shown in Figure 7-2. This form, we believe, will force you to look for as many as five underlying causes for the accident investigated. We also urge your boss to rate your identification of causes (see Exhibit 7-3).

		Circle One	
1. Was it on time?		Yes – 5 pts.	No – 0 pts.
2. Was seriousness indicated?		Yes – 5 pts.	No – 0 pts.
3. Does it say where it happened?		Yes – 5 pts.	No – 0 pts.
4. Can you tell exactly what the injury is?		Yes – 5 pts.	No – 0 pts.

	Circle One
5. How many acts and conditions are listed?	5 4 3 2 1 0
6. How many causes are identified?	5 4 3 2 1 0
7. How many corrections were made or suggested?	5 4 3 2 1 0
8. How many of the listed corrections would have prevented this accident?	5 4 3 2 1 0
9. How many corrections are permanent in nature?	5 4 3 2 1 0
10. In how many of the corrections listed is the supervisor _now_ doing something differently?	5 4 3 2 1 0

Total of Circled Points _____
Multiply X 2 _____

Reviewed by _____ SCORE _____
General Manager

Exhibit 7-3. Investigation rating sheet.

What Should Be Investigated?

An accident that causes death or serious injury, obviously, should be thoroughly investigated. The "near-accident" that might have caused death or serious injury is equally important from the safety standpoint and should be investigated; for example, the breaking of a crane hook or a scaffold rope or an explosion associated with a pressure vessel.

Each investigation should be made as soon after the accident as possible. A delay of only a few hours may permit important evidence to be destroyed or removed, intentionally or unintentionally. Also, the results of the investigation should be made known quickly, as their publicity value is greatly increased by promptness.

Any epidemic of minor injuries demands study. A particle of emery in the eye or a scratch from handling sheet metal may be a very simple case. The immediate cause may be obvious, and the loss of time may not exceed a few minutes. However, if cases of this or any other type occur frequently in the plant or in your department, an investigation might be made to determine the underlying causes.

The Key Facts in Accidents

We have concentrated up to this point on root causes. It could be that in your company additional information will be needed in an investigation. There is a national standard in investigations which you should at least be familiar with.

As explained in American National Standards Institute (ANSI) Standard Z16.2-1995, Information Management for Occupational Safety and Health, the purpose of the standard is to identify certain key facts about each injury and the accident that produced it. These facts are to be recorded on a form which will permit summarization to show general patterns of injury and accident occurrence in as great analytical detail as possible. These patterns are intended to serve as guides to the areas, conditions, and circumstances to which accident prevention efforts may be directed most profitably. Such facts as the following are recorded:

1. **Nature of injury/illness**—the type of physical injury/illness incurred.
2. **Part of body**—the part of the injured person's body directly affected by the injury.
3. **Source of injury/illness**—the object, substance, exposure, or bodily motion that directly produced or inflicted the injury/illness.
4. **Event or exposure**—the manner in which the injury/illness was produced or inflicted by the source of the injury/illness.
5. **Secondary source of injury/illness**—the machine, equipment, object, or substance that either generated the source of the injury/illness or was related to the event or exposure.
6. **Occupation of worker**—uses the Bureau of the Census occupational classification system to describe the kind of work in which the employee is engaged.
7. **Industry of worker**—describes the kind of business in which the employer is engaged according to the Standard Industrial Classification system.
8. **ICD coding**—it is suggested (though not required) that for a more comprehensive recordkeeping and surveillance system, injuries/illnesses also be classified according to the World Health Organization's International Classification of Diseases (ICD).

As you can see, these categories are included in the form in Exhibit 7–1.

Exhibit 7-4 is a checklist of these categories. In addition to unearthing causes in your investigations, your company may want additional information from this checklist so that it can better see patterns of accidents in the company as a whole.

COSTS OF ACCIDENTS

Cost information about accidents is often needed. Some proposed safety corrections may be accepted or rejected on the basis of their effect on profits. While most executives want to make their company a safe place to work, they are also responsible for profits. Usually they are, and should be, reluctant to spend money for accident control unless they can save at least as much as they spend. Without information on the costs of accidents, it is impossible to estimate the savings through accident control.

Work accidents for the purpose of cost analysis fall into two general categories: (1) accidents resulting in work injuries and (2) accidents that cause property damage or interfere with production. No matter what category it falls into or how severe it is, every accident costs a company money. There are two categories of costs: insured and uninsured.

Every organization paying compensation insurance premiums recognizes such expense as part of the costs of accidents. In some cases, medical expenses, too, may be covered by insurance. These costs are definite, and they are known; they are insured costs.

Insured costs can be obtained directly from the insurance company or they can be estimated by a predetermined formula. Your company may use a system for this.

Sometimes referred to as "hidden costs," uninsured costs include such things as:

— Cost of wages paid for time lost by workers who were not injured. These are employees who stopped work to watch or to assist after the accident or to talk about it or who lost time because they needed equipment damaged in the accident or because they needed the output or the aid of the injured worker.

— Cost of damage to material or equipment.

— Cost of wages paid for time lost by the injured worker, other than workmen's compensation payments. (Payments made under workmen's compensation laws for time lost after the waiting period are not included in this element of cost.)

— Extra cost of overtime work necessitated by the accident.

— Cost of wages paid supervisors for time required for activities necessitated by the accident.

1. NATURE OF INJURY

Foreign body	Strain and sprain	Amputation	Dermatitis
Cut	Fracture	Puncture wound	Ganglion
Bruises and contusions	Burns	Hernia	Abrasions
			Others_____

2. PART OF BODY

Head and Neck	**Upper Extremities**	**Body**	**Lower Extremities**
Scalp	Shoulder	Back	Hips
Eyes	Arms (Upper)	Chest	Thigh
Ears	Elbow	Abdomen	Legs
Mouth, teeth	Forearm	Groin	Knee
Neck	Wrist	Others_____	Ankle
Face	Hand		Feet
Skull	Fingers and thumb		Toes
Others_____	Others_____		Others_____

4. ACCIDENT TYPE

Struck against (rough or sharp objects, surfaces etc., exclusive of falls)	Struck by sliding, falling or other moving objects	Overexertion (resulting in strain, hernia. etc.)	Inhalation, absorption, ingestion. poisoning, etc.
Struck by flying objects	Caught in (on or between)	Slip (not a fall)	Contact with electric current
	Fall on same level	Contact with temperature extremes, burns	Others_____
	Fall to different level		

5. HAZARDOUS CONDITION

Improperly or inadequately guarded	Defective tools, equipment, substances	Hazardous arrangement	Poor housekeeping
Unguarded	Unsafe design or construction	Improper illumination	Congested area
		Improper ventilation	Others_____.
		Improper dress	___No unsafe condition

6. AGENCY OF ACCIDENT

Machine	Can and end conveyors (belt, cable, can dividers, chain, twisters, drops, can elevators, etc.)	Hoists and Cranes	Chemicals
Vehicles		Elevators (passenger and freight)	Ladders or scaffolds
Hand tools		Building (door, pillar, wall, window, etc.)	Electrical apparatus
Tin and black plate (sheet, stock, or scrap)	Conveyors (chutes, belt, gravity)	Floors or level surfaces	Boilers, pressure vessels
Material work handled (other than tin and black plate)		Stairs, steps. or platforms	Others_____

8. UNSAFE ACT

Operating without authority	Using equipment, tools, materials or vehicles unsafely	Unsafe loading, placing and mixing	Adjusting, clearing jams, cleaning machinery in motion
Failure to warn or secure		Unsafe lifting and carrying (including insecure grip)	Distracting, teasing
Operating at unsafe speed	Failure to use personal protective equipment	Taking an unsafe position	Poor housekeeping
Making safety devices inoperative	Failure to use equipment provided (except personal protective equipment)		Others_____.
Using defective equipment, materials, tools or vehicles			___No unsafe act

CONTRIBUTING FACTORS

Disregard of instructions	Lack of knowledge or skill	Failure to report to medical department	Others_____.
Bodily defects	Act of other than injured		___No contributing factor

Exhibit 7-4. Checklist for identifying key facts. (From Accident Prevention Manual for Industrial Operations, *7th ed., 1974, Copyright © National Safety Council. Used with permission.)*

DEPARTMENT SUPERVISOR'S ACCIDENT COST REPORT

Injury Accident _____

No-Injury Accident _____

Date _____ Name of injured worker _____

1. How many other workers (not injured) lost time because they were talking, watching, helping at accident? _____

    About how much time did most of them lose ___ hours ___ minutes

2. How many other workers (not injured) lost time because they lacked equipment damaged in the accident or because they needed the output or aid of injured worker? _____

    About how much time did most of them lose _____ hours _____ minutes

3. Describe the damage to material or equipment _____

    Estimate the cost of repair or replacement of above material or equipment $ _____

4. How much time did injured worker lose on day of injury for which he was paid? _____ hours _____ minutes

5. If operations or machines were made idle:
    Will overtime work be necessary to make up lost production?.....................Yes ☐, No ☐.
    Will it be impossible to make up loss of use of machines or equipment?.......Yes ☐, No ☐.

    Contractual penalties or other special non-wage costs due to stopping an operation $ __

6. How much of supervisor's time was used assisting, investigating, reporting, assigning work, training or instructing a substitute, or making other adjustments? _____ hours _____ minutes.

Name of supervisor _____

Fill in and send to the safety department not later than day after accident.

Exhibit 7-5. Department supervisor's accident cost report.

— Wage cost caused by decreased output of injured worker after return to work.

— Cost of learning period of a new worker.

— Uninsured medical cost borne by the company. This cost is usually that of medical services provided at the plant dispensary.

— Cost of time spent by higher supervision and clerical workers on investigations or in the processing of compensation forms.

— Other possible costs are public liability claims, rental fees for equipment, loss of profit on contracts canceled or orders lost if the accident causes a net long-run reduction in total sales, loss of bonuses by the company, wages of new employees if the additional hiring expense is significant, excess spoilage (above normal) by new employees.

Exhibit 7-5 is a form to assist you in estimating the uninsured accident costs. Your company may require you to use such a form.

Other Approaches to Accident Cause Identification

In Chapter 9 some other techniques are described that you can use for cause identification and removal. There are:
— Safety Sampling
— Statistical Safety Control
— Technique of Operations Review
— Incident Recall Technique

All are proven techniques that you may choose to use. For each, the technique is described, an explanation given on how to use it, and suggestions given on how to set objectives, how to measure and reward for the technique.

When your organization holds you accountable for safety performance, they must measure whether or not you are actually doing the activities that are expected of you. If you are expected to do the traditional accident investigation, you will also be measured, as indicated earlier, by things like number of investigations performed, timeliness, quality of the investigation, etc.

INSPECTING FOR HAZARDS

The second core responsibility of a supervisor in a safety program is inspecting for hazards. Inspection is one of the primary tools of safety. At one time, it was virtually the only tool, and it still is the one most used. Many articles have been written about safety inspections, and many have asked the question: "Why inspect?" Some typical answers have been to:

- Check the results against the plan;
- Reawaken interest in safety;
- Reevaluate safety standards;
- Teach safety by example
- Display the supervisor's sincerity about safety;
- Detect and reactivate unfinished business;
- Collect data for meetings;
- Note and act upon unsafe behavior trends;
- Reach firsthand agreement with the responsible parties;
- Improve safety standards;
- Check new facilities;
- Spot conditions.

According to the National Safety Council:

> "The primary purpose of inspection is to detect potential hazards so they can be corrected before an accident occurs. Inspection can determine conditions that need to be corrected or improved to bring operations up to acceptable standards, both from safety and operational standpoints. Secondary purposes are to improve operations and thus to increase efficiency, effectiveness, and profitability."[2]

Systematic inspection is the basic tool for maintaining safe conditions and checking unsafe practices. Each company, plant, or department should develop its own checklist. Sample checklists, stressing work areas or work practices or both, are shown in Exhibits 7-6 through 7-10.

The format of Exhibit 7-6 is typical of many companies' approach. Exhibit 7-7 is similar to this but more detailed. Exhibits 7-8 and 7-9 concentrate on unsafe practices, while Exhibit 7-10 is designed for the concept of multiple causation and symptomatic safety discussed in the last chapter.

Systematic Inspection by Supervisors

Supervisors should continuously make sure that tools, machines, and other department equipment are maintained properly and are safe to use.

To do this effectively, use systematic inspection procedures, and delegate authority to others in your department. For instance, toolroom attendants might inspect all hand tools to see that they are kept in safe condition. Some companies require portable electric tools to be turned in to the electrical department for a monthly check.

Inspection programs should be set up for new equipment, material, and processes. You should see that nothing is put into regular operation until it has been checked for hazards, its operation studied, additional safety devices installed if necessary, and safety instructions developed.

You might give your lead men the responsibility for inspecting equipment and for seeing that their workers observe safe practices. If you do, make certain that all inspections by persons other than you are up to your standards. Spot check periodically to make sure assignments are being carried out, that safety precautions are being observed, and that equipment is running efficiently and safely.

Inspection of Work Practices

You should regularly observe your people to determine whether they are working in the safest way and complying with your rules and wishes. As

INSPECTING FOR HAZARDS

FOREMAN'S INSPECTION FORM			
Name _____			Date _____
Item	Good	Poor	Disposition
Housekeeping			
Aisles _____			
Piling _____			
Floor surfaces _____			
Tools			
Condition _____			
Grounding _____			
Guards _____			
Personal protection _____			
Miscellaneous _____			
Ladders _____			
Slings _____			

Exhibit 7-6. Typical inspection form.

you look over your area keep in mind such questions as those in the list that follows.

— Do employees operate machinery or use tools, appliances, or other equipment without authority?

— Are they working or operating at unsafe speeds?

— Have guards been removed, or have guards or other safety devices been rendered ineffective?

— Do people use defective tools or equipment, or use tools or equipment in unsafe ways, or use hands or body instead of tools?

SAFETY INSPECTION CHECK LIST

Plant or Department _____ Date _____

This list is intended only as a reminder. Look for other unsafe acts and conditions, and then report them so that corrective action can be taken. Note particularly whether unsafe acts or conditions that have caused accidents have been corrected. Note also whether potential accident causes, marked "X" on previous inspection, have been corrected.

(✔) indicates *Satisfactory*   (X) indicates *Unsatisfactory*

1. FIRE PROTECTION
- Extinguishing equipment .... ☐
- Standpipes, hoses, sprinkler heads and valves .... ☐
- Exits, stairs and signs .... ☐
- Storage of flammable material ..... ☐

2. HOUSEKEEPING
- Aisles, stairs and floors .... ☐
- Storage and piling of material ..... ☐
- Wash and locker rooms .... ☐
- Light and ventilation .... ☐
- Disposal of waste .... ☐
- Yards and parking lots .... ☐

3. TOOLS
- Power tools, wiring .... ☐
- Hand tools .... ☐
- Use and storage of tools .... ☐

4. PERSONAL PROTECTIVE EQUIPMENT
- Goggles or face shields .... ☐
- Safety shoes .... ☐
- Gloves .... ☐
- Respirators or gas masks .... ☐
- Protective clothing .... ☐

5. MATERIAL HANDLING EQUIPMENT
- Power trucks, hand trucks .... ☐
- Elevators .... ☐
- Cranes and hoists .... ☐
- Conveyors .... ☐
- Cables, ropes, chains, slings ...... ☐

6. BULLETIN BOARDS
- Neat and attractive .... ☐
- Display changed regularly .... ☐
- Well illuminated .... ☐

7. MACHINERY
- Point of operation guards .... ☐
- Belts, pulleys, gears, shafts, etc. ... ☐
- Oiling, cleaning and adjusting ..... ☐
- Maintenance and oil leakage ...... ☐

8. PRESSURE EQUIPMENT
- Steam equipment .... ☐
- Air receivers and compressors .... ☐
- Gas cylinders and hose .... ☐

9. UNSAFE PRACTICES
- Excessive speed of vehicles ...... ☐
- Improper lifting .... ☐
- Smoking in danger areas .... ☐
- Horseplay .... ☐
- Running in aisles or on stairs ..... ☐
- Improper use of air hoses .... ☐
- Removing machine or other guards . ☐
- Work on unguarded moving machinery .... ☐

10. FIRST AID
- First aid kits and rooms .... ☐
- Stretchers and fire blankets ...... ☐
- Emergency showers .... ☐
- All injuries reported .... ☐

11. MISCELLANEOUS
- Acids and caustics .... ☐
- New processes, chemicals and solvents .... ☐
- Dusts, vapors, or fumes .... ☐
- Ladders and scaffolds .... ☐

Signed _____

USE REVERSE SIDE FOR DETAILED COMMENTS OR RECOMMENDATIONS

Exhibit 7-7. Safety inspection check list.

SUMMARY OF UNSAFE PRACTICES

Area or Division _____ Force _____

District _____ Period Covered _____

	GROUP (FORCE, DISTRICT, DIVISION OR AREA)					TOTAL
1	SUPERVISORS REPORTING UNSAFE PRACTICES					
2	SUPERVISORS NOT REPORTING UNSAFE PRACTICES					
3	TOTAL NO. OF SUPERVISORS					
4	TOTAL NO. OF EMPLOYEES IN GROUP					
5	NUMBER OF UNSAFE PRACTICES REPORTED					

CAUSE OF UNSAFE PRACTICES

6	SUPERVISION	LACK OF ANALYZING, OR PLANNING THE WORK					
7		INADEQUATE TASK TRAINING					
8		LACK OF DEFINITE OR SPECIFIC INSTRUCTIONS					
9		IMPROPER ASSIGNMENT OF EMPLOYEE					
10		FAILURE TO SEE THAT INSTRUCTIONS WERE FOLLOWED					
11		OTHER					
12	EMPLOYEE	LACK OF ANALYZING, OR PLANNING THE WORK					
13		DISREGARD FOR KNOWN SAFE PRACTICES					
14		LACK OF EXPERIENCE					
15		ABSTRACTION OR FORGETFULNESS					
16		HASTE					
17		OTHER					
18	TOTAL						

UNSAFE PRACTICES

19	MOTOR VEHICLES—OPERATION AND MAINTENANCE					
20	POLES—WORKING ALOFT					
21	ACTION, BOTH IN AND OUT OF BUILDINGS THAT MIGHT RESULT IN SLIPS OR FALLS					
22	LADDERS—EXTENSION, STEPLADDERS, AND MOBILE PLATFORMS					
23	BODY HARNESSES AND LANYARDS					
24	CLIMBERS, PADS, AND STRAPS					
25	GUARDING EMPLOYEES AND PUBLIC					
26	TOOLS AND MATERIALS					
27	GOGGLES					
28	USE OF RUBBER GLOVES AND OTHER PROTECTIVE DEVICES AND PRECAUTIONS TAKEN AROUND LIVE WIRES					
29	MANHOLES, CONDUIT AND EXCAVATIONS					
30	FIRST AID FOR AND CARE OF INJURIES					
31	MISCELLANEOUS					
32	TOTAL					

RECOMMENDATIONS _____

Observed by _____

Title _____

Exhibit 7-8. Summary of unsafe practices. (From Accident Prevention Manual for Industrial Operations, *7th ed., 1974, Copyright © National Safety Council. Used with permission.)*

(See Note under "UNSAFE PRACTICES" on other side of page.)

	REF.	NO.		REF.	NO.
19. MOTOR VEHICLES • OPERATION AND MAINTENANCE			**26. TOOLS AND MATERIALS**		
Improper Parking			Improper Use of Tools and Materials		
Failure to Conform to Traffic Laws			Use of Defective or Unauthorized Tools		
Stepped Off or On Vehicle in Motion			Tossing Tools and Materials		
Backing Without Taking Proper Precautions			Dropping Tools and Materials from Aloft		
Poor Housekeeping on Truck or Car			Unsafe Lifting or Handling		
Lights, Brakes, Horn, etc. Not Tested			Unsafe Carrying of Tools, Nails, Tacks, etc.		
Unsafe Handling of Derrick			Improperly Stored or Placed		
Under Derrick in Operation			Unsafe Use of Cable Car		
Hands on or Near Winch Line Sheave or Drum			Handline Dangling		
Unsafe Handling and Use of Trailer			Tools and Materials Left Lying Around		
Unsafe Performance of Maintenance Operations			Failure to Use Tree Sling		
			Unsafe Cutting of Wire—Flying Ends or Pieces		
			Not Using Flashlight Where Required		
20. POLES • WORKING ALOFT					
Failure to Test or Make Safe Before Climbing			**27. GOGGLES • FAILURE TO USE WHEN:**		
Failure to Compensate for Unbalanced Load			Drilling Concrete brick or Other Masonry		
Working in Dangerous Position			Grinding, Chipping, Handling Brush, etc.		
Lack of Care in Climbing					
Standing Under Workman Aloft			**28. USE OF RUBBER GLOVES, ETC., AND PRECAUTIONS TAKEN AROUND LIVE WIRES**		
Lack of Care Approaching or Leaving Pole			Failure to Use Rubber Gloves when Required		
21. ACTION BOTH IN AND OUT OF BUILDINGS, THAT MIGHT RESULT IN SLIPS OR FALLS			Failure to Test and Maintain Properly		
			Working Too Close, Live Wires When Aloft		
Unnecessary Running, Stairs and Across Floors			Lack of Precautions to Prevent Wire From	X X X	X X X
Lack of Ordinary Care Outside Building			Flipping Up Into Live Wires		
Sitting on Tilted Chairs			Sagging or Falling Into Live Wires		
Standing on Chairs, Boxes, Cans, etc.			Reel Tender Not Protected		
			Unsafe Use of Chain Hoist, Tent, Steel Tape, etc.		
22. LADDERS • EXTENSIONS, MOBILE AND STEP LADDERS			Incomplete Survey After Suspected E. L. Contact		
			Carelessness Around Electric Circuits and Equipment		
Failure to Make Secure (Footing Lashing, Holding)			Failure to Use Adequate Protective Devices		
Failure to Place Leg Thru Ladder or Use Safety Pull Rope Not Tied					
Defective Ladder, Spurs Not Turned, etc.			**29. MANHOLES, CONDUIT AND EXCAVATIONS**		
Overreaching Too High Up on Ladder, etc.			Failure, Proper Tests for Gas or Oxygen Deficiency		
Improper Angle for Climbing			Insufficient Ventilation (with Sail or Blower)		
Failure to Inspect			Entered Manhole Without Manhole Guard		
Unguarded at Hazardous Location			Failure to Use Ladder in Manhole or Excavation		
Failure to use Where Required or Used Wrong Kind			Insufficient Guarding, Manholes or Excavations		
Not Lashed to Motor Vehicle or Carried Properly			Unsafe Removal and Handling of Cover		
Failure to Extend Spreaders of Stepladder			Smoking or Open Flames Near or In Manholes		
Tools or Material Left on Steps or Top					
Shoved Mobile Ladder Endangering Person on Ladder			**30. FIRST AID FOR AND CARE OF INJURIES**		
			Failure to Give Proper First Aid		
23. BODY BELTS AND SAFETY STRAPS			Did Not Continue Proper Care of Injury		
Failure to Use While Aloft (On Pole, Platform, etc.)			Failure to Report Injury		
Failure to Look, Feel and Know Snaphook is Secure			First Aid Kit Not Properly Maintained		
Not Inspected or Properly Maintained					
			31. MISCELLANEOUS		
24. CLIMBERS, PADS AND STRAPS			Horseplay On The Job		
Wearing in Trees, Vehicles, on Ladder, Ground, etc.			Debris Left Lying Around		
Failure to Inspect and Maintain Properly			Unsafe Position on St., Hwy, Rd., R.R. Track, etc.		
Unsafe Climbing Habits			Handling of Hot Solder, Paraffin, etc.		
			Poison Ivy or Oak—Lack of Protection or Care		
25. GUARDING EMPLOYEES AND PUBLIC			Clothing and Shoes—Poor, Insufficient		
Failure to Use Adequate Warning Signs or Flags			Trees or Brush—Cutting and Handling		
Keeping Children and Others Away from Operation					
Guarding Public at Dangerous Location					

Exhibit 7-9. Unsafe practices observed. *(From* Accident Prevention Manual for Industrial Operations, *7th ed., 1974, Copyright © National Safety Council. Used with permission.)*

FOREMAN'S INSPECTION FORM		
Name _____ Date _____		
Symptom noted **Act/Condition/Problem**	**Causes** **Why — What's Wrong**	**Corrections made or suggested** **By you — By others**

Exhibit 7-10. Suggested inspection form. (From Techniques of Safety Management *by Dan Petersen. Copyright © 1987 by Dan Petersen.)*

— Do they overload, crowd, arrange, or handle objects or materials unsafely?

— Do people stand or work under suspended loads, open hatches, shafts, or scaffolds; or ride loads; or get on or off equipment or vehicles in motion; or walk on railroad tracks; or cross car tracks or vehicular thoroughfares except at crossings?

— Do they repair or adjust equipment in motion, under pressure, electrically charged, or containing dangerous substances?

— Does anyone or anything distract the attention of or startle workers?

— Do the workers fail to use safety equipment, or do they use inadequate, badly adapted, or wrongly chosen personal protective equipment or safety devices?

— Do the employees have poor housekeeping habits or fail to remedy unsanitary or unhealthful conditions?

— Is horseplay activity a common occurrence?

Your regular inspections should also focus on ergonomics. Ergonomics is looking at how your machines, systems and procedures interact with your people. For instance, many people in their normal work must use parts of their body repetitively. Over time, this can result in back problems, arm and shoulder pains, wrist injuries, etc. Much of this can be reduced through an ergonomic analysis.

This ergonomic analysis has as its primary focus the preventing of repetitive motion injuries, which are commonly called Cumulative Trauma Disorders (CTDs).

CTDs are occupational injuries that develop over time, affecting the musculoskeletal and peripheral nervous systems. They can develop in any part of the body, but are most prevalent in the arms and back. These injuries are caused by jobs that require repeated exertions and movements near the limits of the individual's strength and range-of-motion capability. These movements, although not initially painful, cause microtraumas to the soft tissues. Over time, small strains to the muscle/tendon/ligament system build up, resulting in fatigue and soreness. If the individual continues the action that is causing the pain, cumulative trauma disorders are likely to develop. CTDs can have the following affects on individuals:

- pain
- numbness or loss of sensation
- reduced strength
- degraded ability to perform work
- degraded ability to participate in leisure activities

Other Approaches to Hazard Identification

Other approaches, besides inspection, are spelled out later:
— Job Safety Analysis
— Hazard Hunts
— OSHA Compliance Checks
— Ergonomic Analysis

Each is described, should you choose to use the technique. How objectives can be set and how your performance can be measured is also spelled out.

Should you choose to do straight inspections as described in this chapter, objectives and measures like number of inspections, numbers of hazards removed, etc., can be used in your company's accountability system.

COACHING

The third key responsibility is coaching. We use the term coaching instead of training because coaching connotes a wider range of activities than training does. We are interested in all activities you should consider to improve specific job performance, and not in a dry, classroom approach that the term training often brings to mind.

To coach means to help a person do better. Any supervisor will admit doing this. Some organizations have extensive training programs, and others have very limited training. In either case, you have to be a coach on your job. Any supervisor will readily admit having a responsibility to help his people do better.

For our purposes, we'll say that coaching consists of doing these kinds of things: explaining, demonstrating, correcting, encouraging, and reprimanding. Putting the coaching process in its simplest form, we can say that it has three steps: (1) finding out where the employee is now; (2) finding out where we want the person to be; and (3) providing the difference.

Unfortunately, in safety we usually spend almost all our time determining content and method. Theorists tell us to spend the bulk of our effort and concentration on (1) and (2) above. If we do a good job in analyzing these, the third step usually falls into place naturally. Theorists tell us that content is no more than the difference between the first two steps and that the method of presentation is almost immaterial to learning.

Every supervisor has two divergent coaching tasks: To coach each person individually and to coach the group collectively.

The Individuals

The supervisor must look at each person individually and go through the process described above of finding out the person's needs—where the person is now in comparison to where you want him to be. Then with explanations, demonstrations, corrections, encouragements, and reprimands, each individual must be coached to achieve the performance level you want.

When should the individual be coached? Obviously, coaching should occur whenever you feel the person needs to do better in any area. Coaching seems to be particularly appropriate in the following kinds of situations:

When a new person is assigned to you. You need to show him around, introduce him to people, explain the purpose of your department, instruct him on safety rules and regulations, point out special hazards, and in other ways get him oriented to the job he is to do. This is a task that you are already doing. By giving it conscious attention, you'll do it better and improve the person's performance.

When you see an unsafe act. Often a word or comment or brief explanation is sufficient to correct it. Sometimes there is a need for greater explanation or demonstration. At the back of our minds is the thought that an unsafe act continued long enough will finally produce an accident. Safety is not a goal that can be measured in daily or hourly figures. Your people know safety is important by what you do in your daily and hourly contacts with them. When you turn your back on unsafe acts, when you fail to do the coaching they call for, they quite readily accept the fact that safety is not an important goal or important in their work.

When we see anything well done. It is easier to improve performance by praise of the job well done than it is by criticism. Individuals react positively to compliments and want to continue to receive them.

When we give orders or work assignments. Our orders should clearly express the fact that we want something. This situation is a red flag that should alert us to give additional coaching at this time.

When we see anything being done incorrectly. If we coach when we see things well done, our people are more ready to accept our coaching when we see something done incorrectly. When we praise, we establish what we do want. When we see anything done incorrectly, we can both establish what it is that we do not want and use the opportunity to establish what we do want.

Whenever we assign an unusual job. Much of our work falls into rather usual patterns, but in every organization there are some situations that are unusual. Every unusual situation or unusual job suggests to us that we have a need to coach, to explain, to demonstrate, to make sure that people know what we want, for it is in these circumstances that we can expect a severe injury.

The Group

The supervisor must also look at the department as a whole and assess its needs. And he should look for occasions or situations when this can be most effective. Some of these are:

When you see a better way to do things.

When new products, methods, or machines are introduced. Even though only some of your people will operate the machines, it is well to acquaint the whole group with the new product, method, or machine. Let all understand how it fits into the total group effort.

When unsafe conditions crop up. An unsafe act is performed by an individual. But unsafe conditions require a group effort. You cannot see or be present every time an unsafe condition crops up. They must be seen and corrected or reported as a group effort.

When establishing your goals. Supervisors often do not clearly establish and explain their goals to their people. Usually you have a general idea of an end result desired by the work of the whole group and a general idea of the performance required and desired by the whole group. This needs to be explained and discussed with your group. The supervisor's job is to get the group to want to achieve his goals or (more precisely) to cause the supervisor's goals to become the goals of the group.

When any unusual condition affects the whole group. This may come about as the result of an accident, a slow-down in work with necessary layoffs, an unusual or special job, or any occasion in which the understanding of the whole group and the cooperation of the whole group is important to accomplish the end desired.

How to Coach

While we won't discuss specific methods here, we do want to bring up a few ideas on human learning. There has been more research on this topic than on almost any other in psychology. Here are some of the "knowns":

Motivation and learning—Learning theorists agree that an individual will learn best if he is motivated toward some goal that is attainable by learning the subject matter presented. The behavior of people is oriented toward relevant goals, whether these goals are safety, increased recognition, production, or simply socializing. People attempt to achieve those goals that are important to them at the moment, regardless of what's important to you.

Reinforcement and learning—Positive rewards for certain behavior increase the probability that the particular behavior will occur again. Negative rewards decrease that probability. Reinforcement is most important to learning.

Practice and learning—An individual learns best through practice and involvement.

Feedback and learning—Telling a worker how he is doing is essential for learning. It is difficult for the worker to improve his performance unless he is given some knowledge of his performance. What are his errors? How can he correct them? This is essential.

Meaningfulness and learning—In general, meaningful material is learned better than material that is not meaningful. In order to simplify the worker's learning task, make the material as meaningful to him personally as possible.

Later we'll discuss a number of different ways you may wish to consider, ways to coach. Safety meetings are only one way—there are a number of others that have been found to be much more effective. One on one contacts typically get much better results, for example.

In addition, the following techniques will described:
— Job Safety Observations
— Safe Behavior Reinforcement
— One-Minute Safety Program
— Stress Assessment Technique

For each objective, setting and measurements will be spelled out.

MOTIVATING

The final key responsibility is to motivate your people. The next chapter discusses this in some detail in terms of what turns people on (and off) today.

The Influence of Management

How do you motivate the worker to be safe? No one has the magic key to understanding all people—and to knowing how to get them to want to do what needs to be done. All we can do is attempt to gain some insights. We'll look at what influences you have as a supervisor. These influences help to mold and shape the employee's decision about how he will work. He or she must make this decision, management cannot. But you can place considerable pressure on him or her.

For example, let us imagine a factory worker named Jim, who slaves over a hot machine all day to produce 275 Super Speed Fishing Worm Untanglers, Size 4. One day management hands out booklets telling Jim and his fellow workers how they can produce 300 untanglers instead of 275. Nobody in his right mind enjoys making fishing worm untanglers. So Jim hastily skims through the booklet, throws it away, and keeps right on turning out 275 a day.

Management, through you, then sends down an order warning that any employee who fails to produce 300 untanglers a day will be fired. Jim is now powerfully motivated. He finds that the booklet is interesting reading after all, and he learns everything in it with remarkable ease. He also produces 300 untanglers. You have thus extended an influence that has motivated Jim. You have not only changed his production level, you have also made him interested in booklets.

Management, through its policy, makes the decision that safe performance from employees is desirable. Management cannot, however, force a safe performance. Each employee decides for himself whether or not he will work, how hard he will work, and how safely he will work. His attitudes shape his decision—attitudes toward himself, his environment, his boss, his company, his entire situation. His decision is based on his knowledge, his skills, and his group's attitude toward the problem. All that you in supervision can do is to recognize those influences over which you have control and extend your influence wherever possible.

The Influence of Supervisors

Practically all your company's motivational attempts reach employees go through their supervisors. You are, among other things, a funnel, directing all material and information to the employee. You also direct or carry out the vast majority of training. Everything that motivates employees is applied by you. Obviously, your role is crucial. According to the National Safety Council:

> The supervisor is the key man in any program to create and maintain interest in safety because he is responsible for translating management's policies into action and for promoting safety activities directly among the employees. How well he meets this responsibility will determine to a large extent how favorably the employees receive the safety activities.
>
> The supervisor's attitude toward safety is a significant factor in the success, not only of specific promotional activities, but also of the entire safety program, because his views will be reflected by the employees in his department.
>
> The supervisor who is sincere and enthusiastic about accident prevention can do more than the safety director to maintain interest.
>
> Conversely, if the supervisor pays only lip service to the program or ridicules any part of it, his attitude offsets any good that might be done by the safety professional.
>
> Setting a good example, for instance, wearing safety glasses and other personal protective equipment whenever it is required, is one of the most effective ways in which the supervisor can promote safety.

Teaching safety is an important function of the supervisor. He cannot depend upon safety posters, a few warning signs, or even general rules to do his job of training and supervision. A good balance of basic training and supervision and judicious use of promotional material proves effective. However, the supervisor himself must first be trained if he is to be competent.

Supervisors can be most effective in giving facts and personal reminders on safety to employees. This procedure is particularly necessary in the transportation and utility industries, where crews are on their own from terminal to terminal.

In any case, supervisors should be encouraged to take every opportunity to exchange ideas on accident prevention with workers, to commend them for their efforts to do the job safely, and to invite them to submit safety suggestions.

Your role is indeed crucial to the success of your company's safety results. And your success in carrying out this role is dependent almost entirely on your skills in dealing with your people. You cannot succeed in safety or in anything else without people. You need them. You need to understand them, and they need to understand you. Chapter 8 looks at people and your interpersonal relationships with them.

Notes

1. From *Industrial Accident Prevention* by H. W. Heinrich. Copyright 1969 by Viola E. Heinrich. Used with permission of McGraw-Hill Book Company.
2. Gary R. Krieger and John F. Montgomery, eds., *Accident Prevention Manual for Business and Industry, Administration & Programs*, 11th ed. (Itasca, Illinois: National Safety Council, 1997).

CHAPTER 8 The Dynamics of Supervision

UNDERSTANDING WORKERS

In the last chapter it might be inferred (improperly) that we as supervisors can "motivate" our people. This is really not true—you cannot actually "motivate" anyone. Motivation comes entirely from within—an employee motivates himself. Our definition of motivation is "desire by an employee to want to do what you want him to." His wanting the same goal you want is the key, for then you've no problems in getting him to act in the manner and direction you want.

As we indicated before, all you can do is to influence his decisions by creating an atmosphere in which he can decide comfortably to want to go your way. And the only way you can effectively do this is to understand him—his needs, his wants, his way of thinking.

THE SINS OF SUPERVISION

Knowing and understanding your workers is important, but it is also important to know yourself and how you handle your job. Poor supervisory practices can produce negative results just as poor attitudes and lack of awareness can. Do you recognize any of your tactics in this summary of what turns employees off?[1]

Using straight-jacket controls. Through lack of confidence in his own staff, a supervisor may install expensive, time-consuming systems that he hopes will make his operation foolproof. Instead. all the constricting regulations, restraints, records, and checks interfere with both work performance and the exercise of judgment by employees. Employees have to spend too much of their time checking and double-checking petty details.

Their initiative is bound to be stifled in the dense undergrowth of rules.

Being unclear. Giving vague orders frequently causes costly misunderstandings—misunderstandings that could have been avoided if the employee had been told clearly what he was supposed to do.

Being inconsistent. Frequent arbitrary changes in work rules and department policies can be murder on employee morale. An employee should be able to feel that the standards for judging his work won't be changed from day to day.

Ignoring employee potential. The supervisor who is too busy to show any concern for the goals of his subordinates won't make much headway in achieving his own goals.

Playing favorites. A supervisor can't treat all his subordinates alike, but he must be impersonal in his approach.

Being too conservative. Wearing blinders to ward off new ideas is one way to ward off a progressive, dynamic department, too. Some supervisors get nervous when a subordinate suggests a fresh way of doing things—they know the old way works, so why take chances on an untried method? Once this reluctance to consider new ideas becomes known throughout the department, the supervisor can relax—he won't be offered any. Who wants to make the boss unhappy?

Abusing status privileges. A supervisor has a special status of which he can be justly proud. He may have certain privileges that he has earned by his promotion to managerial ranks. But if he misuses these privileges, he is betraying the responsibilities of his job.

INCONGRUENCY THEORY

Let's examine some of the reasons behind a few of our safety and other management problems as explained by Chris Argyris of Yale University.[2] Argyris suggested that the problems of worker apathy and lack of effort are not simply a matter of individual laziness. Rather they are often healthy reactions by normal people to an unhealthy environment created by common management policies. More specifically, Argyris states that most adults are motivated to be responsible, self-reliant, and independent. But the typical business organization confines most of its employees to roles that provide little opportunity for them to act in this manner.

Argyris proposed and tested a theory (called the Incongruency Theory) which provides supervisors some insights into the reasons why humans commit errors. First, he looked at the nature of man as he develops from an infant into an adult. As a child the human is passive and dependent on his parents. He exhibits few behaviors; his interests are shallow and short term. He is at all times a subordinate in his relationships with his parents and is relatively lacking in self-awareness. As he matures this changes.

The mature adult is active and is an independent creature who likes to stand on his own. He exhibits many behaviors, and his interests are deep and long term in nature. He views himself as an equal in most relationships, not as a subordinate, and he is self-aware. This evolution is what maturation is all about.

Argyris then looked at the characteristics of organizations. In his view all organizations, whether industrial, governmental, mercantile, religious, or educational, are structured according to certain principles:

Chain of command—this creates superior-subordinate relationships. It creates dependency on the boss, passivity on the part of the worker, and shorter and shallower interests on the part of the worker.

Short span of control—this creates dependency, reduces the freedom and independence of the worker.

Unity of command—there is only one boss, which again creates dependency and heightens the subordinate role of the worker.

Specialization—the work should be broken down into small simple tasks and each task assigned to a separate worker. This creates shorter, shallower interests, a lack of self-fulfillment and self-importance, fewer behaviors, dependency, and passivity.

A basic conflict exists between the characteristics of the mature worker and the characteristics of the organization he works for. They are pulling at cross-purposes. Yet this seems to be inevitable.

Argyris suggests this conflict causes people to quit (turnover), to quit mentally (apathy), to lose motivation and interest in the company and its goals, to form informal groups, to cling to the group norms instead of company established norms, and to evolve a psychological set against the company, believing it is wrong in most things it attempts to do. These factors also cause accidents because of inattention, disregard of safety rules, and a poor "attitude" toward the company and safety.

The normal management reaction to these symptoms is more control, more specialization, more pressure. Management believes even more strongly that "they" (the workers) must be controlled (treated like children). The problem becomes circular. If workers are treated as if they are children, they begin to act like immature people. What can we do about it? Since we cannot feasibly change mature people into immature ones (nor would we want to) the only option is to look at the organization and see how its characteristics can be changed. This leads us to organizations (and safety programs) with less control, less specialization, and less superior-subordinate relations and to thoughts about participative leadership and a concept known as job enrichment. These are discussed next.

WHAT TURNS THEM ON

After considering how to turn employees off, let's look at how to turn them on. First, a look at the young again and then a look at you, the supervisor, and what you can do.

A lot of research has been done in recent years on what effective supervisors do that is different from what ineffective supervisors do. One researcher, Dr. Rensis Likert of the University of Michigan, found that supervisors with the best performance focus their primary attention on the human aspects of their subordinates' problems and on building effective work groups with high performance goals. These supervisors are referred to as employee-centered leaders. (Job-centered managers are concerned more with completing tasks than with how the work routine affects the employees.) Exhibit 8-1 presents the findings from this study.[3] It indicates that employee-centered supervisors tend to have better productivity records than job-centered supervisors.

Dr. Likert's studies also found a marked inverse relationship between the amount of unreasonable pressure the workers feel and the productivity of the department. (See Exhibit 8-2.) Feeling a high degree of pressure is associated with low performance. General rather than close supervision is associated with high productivity. This relationship is shown in Exhibit 8-3. Supervisors in charge of low-producing units tend to spend more time with their subordinates than do the high-producing supervisors, but the time is broken into many short periods in which they give specific instructions, "Do this, do that, do it this way."

Genuine interest on the part of a superior in the success and well-being of his subordinates has a marked effect on their performance, according to Dr. Likert. Exhibit 8-4 shows that high-producing foremen tend either to ignore the mistakes their subordinates make, knowing that they have learned from the experience, or to use these situations as educational experiences by showing how to do the job correctly. The foremen of the low-producing sections, on the other hand, tend to be critical and punitive when their subordinates make mistakes. Thus, one way to turn employees on is to be supportive, to set high goals, to avoid "bugging" them with close supervision, and to limit criticism of performance.

Theory X Versus Theory Y

Different authors use different terms, different classifications, and different labels for various types of supervisors. Douglas McGregor used as his basis of classification some basic assumptions the supervisor makes about the nature of man.[4] On the one hand, according to McGregor, is the The-

	Job-Centered	Employee Centered
High-producing sections	1	6
Low-producing sections	7	3

Exhibit 8-1. Productivity of employee-centered supervisors vs. job-centered supervisors.

	Department Productivity	
	Below Average	Above Average
The ten departments which feel the least pressure	1	9
The middle eleven departments	6	5
The ten departments which feel the most pressure	9	1

Exhibit 8-2. Relationship between pressure felt by employees and their production.

	Number of First-Line Supervisors Who Are:	
	Under close supervision	Under general supervision
High-producing sections	1	9
Low-producing sections	8	4

Exhibit 8-3. Relationship between closeness of supervision and production.

	Foremen's Reaction to a Poor Job (As Reported by Their Men)	
	Punitive: Critical	Nonpunitive: Helpful
High-producing foremen	40%	60%
Low-producing foremen	57%	43%

Exhibit 8-4. Relationship between supervisor's reaction to a poor job and production.

ory X supervisor who holds the basic assumption that people do not like to work and must be forced or coerced into it. On the other hand is the Theory Y supervisor who believes that people innately like to work and achieve. Obviously, the management styles of supervisor X and supervisor Y will be quite different because of these assumptions.

Supervisor X will check and double-check his employees to make sure first that they are actually on the job and second that they are working.

He will build many controls to ensure that the jobs he has responsibility for are being performed properly. Supervisor Y will spend his time removing barriers between his employees and their work. These barriers could be production delays, foul-ups, or company policies and rules which he does not agree with. He will try to do whatever is necessary for his employees so that they can concentrate on the work tasks at hand. Once this is done, his workers will achieve as they want to, he believes, and the job he is responsible for will get done. Management thinking today is that the Theory Y supervisor is more successful.

DUAL FACTOR THEORY

While Chris Argyris and other behavioral scientists have come to the conclusion that job enrichment is a necessity, it was actually Frederick Herzberg of Case-Western Reserve University who first coined the term and expressed the principle.[5] Herzberg's interest in job enrichment grew out of his discovery of what might be called the dual factor theory of job satisfaction and motivation.

The essence of Herzberg's theory can be illustrated by comparing the "traditional" way of viewing dissatisfaction and motivation as simple opposites on a single line with his view which places motivation on a separate and distinct line from dissatisfaction (see Exhibit 8-5). The classical approach to motivation has concerned itself only with the environment in which the employee works, that is, the circumstances that surround him while he works and the things he is given in exchange for his work. Herzberg considers this concern with the environment a never-ending necessity for management, but says it is not sufficient in itself for effective motivation. That requires consideration of another set of factors, namely, experiences inherent in the work itself.

Herzberg asserts that work itself can be a motivator. Traditionally work has been regarded as an unpleasant necessity rather than a motivation. He suggests that these kinds of things are motivators (in this order): achievement, recognition, work itself, responsibility, and growth. And that these kinds of things are dissatisfiers: company policies and administration,

```
TRADITIONAL
├─────────────────────────────┤
Dissatisfaction                Motivation

MOTIVATION-HYGIENE
├─────────────────────────────┤
Dissatisfaction                        No
                                  satisfaction
├─────────────────────────────┤
Unmotivated                      Motivated
```

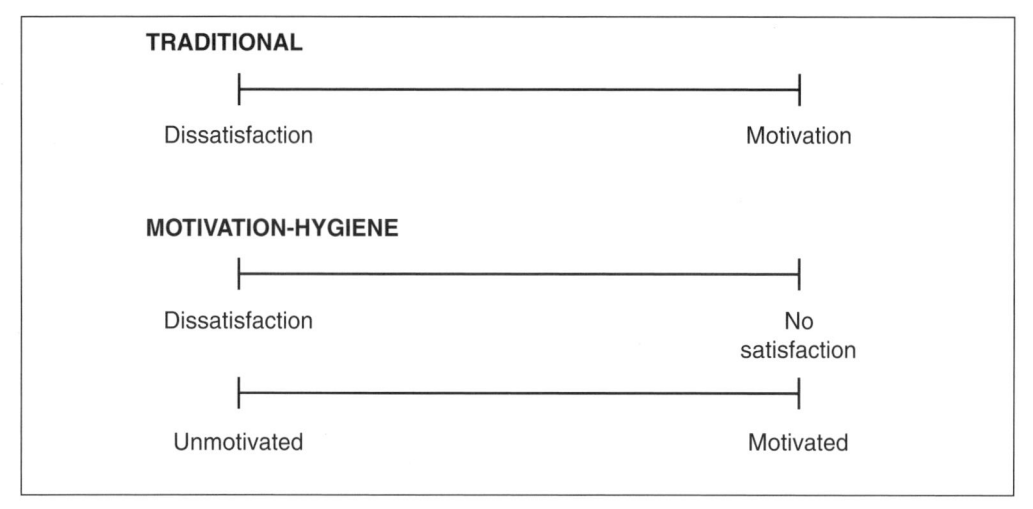

Exhibit 8-5. The dual factor theory.

supervision, working conditions, interpersonal relations, and money, status, and security. In other words, the factors in the work situation which motivate employees are different from the factors that dissatisfy employees. Motivation stems from the challenge of the job while dissatisfactions more often spring from factors peripheral to the job.

Turning Them On

After reading about the various theories of management, you probably have a good idea of what turns on your workers. Let us summarize a few of the kinds of practices that motivate your employees to have the same goals you want.

1. Being employee-centered (a person who is supportive, sets high goals, does not "bug" them, and is not overly critical).
2. Using Theory Y assumptions and Theory Y leadership styles.
3. Providing a job that has meaning.

ATTITUDES

We are concerned in safety work with job behavior; job behavior is governed by a worker's attitude. Our attitudes are the result of our own personal experiences, and their emotional content often is more important than the facts involved.

Attitude and Safety

In his book, *Supervisor's Guide to Human Relations*, Dr. Earle Hannaford defines attitude as the potential for action and safety attitude as "a readiness to respond effectively and safely, particularly in tension-producing situations." He goes on to state the three components of attitude—

FOUR STEPS IN ATTITUDE FORMATION	TYPICAL SAFETY ACTIVITIES TO USE
Step I Laying the Foundation for the Attitude	**Mass Media** Safety slogans, safety posters, safety talks. Motion pictures and sound strip films of general safety nature. Training classes and demonstrations for groups on job methods and theory. Company safety policies. Safety contests and competitions of a group or company nature.
Step II Personalizing the attitude for the individual	**Learned Responses and Habit Formation** On-the-job training in correct safe work methods. Good supervision—immediate correction of violations of safe working practices to build safe habits. Individual participation in safety meetings, safety planning and safety inspections. Motion pictures and sound strip films dealing with job methods and sequences. Recognition of personal contributions by boss and higher authority figures. Individual safety awards.
Step III Fixation of the Attitude Emotional Set	**Emotional Set** Discussion of actual job-related accidents with individual participation. Role playing—permits identification with and projection of self by individual. Motion pictures with high emotional content relating to safety in general and to job performance. Actual demonstration of their personal interest in safety by the boss and higher management—making it the No. 1 item—catching the attitude from authority figures.
Step IV Keeping the attitude alive.	**Attention, Memory and Emotional Set** Checkup on attitude status of individuals and groups using industrial safety attitude scales for employees and supervisors to see where emphasis is needed. Attitude surveys.

NOTE: Plan safety program to offset "safety program fatigue" by using some of the items designed to provide for Steps I, II, and III since employees may be in any one of these steps or may have regressed from III to II, or II to I.

Exhibit 8-6. Using practical safety program activities to build good safety attitudes. *(From* Supervisor's Guide to Human Relations *by Earle Hannaford. Copyright © 1967 National Safety Council, used with permission.)*

Attitude = learned responses + habit + emotional set

and suggests there are four steps in building attitudes[6] (see Exhibit 8-6).

Dr. Hannaford's work with attitudes has shown the relationship between attitudes and results in the area of safety (see Exhibit 8-7, A and B). As is readily evident in Exhibit 8-7A, the poorer the safety attitude of

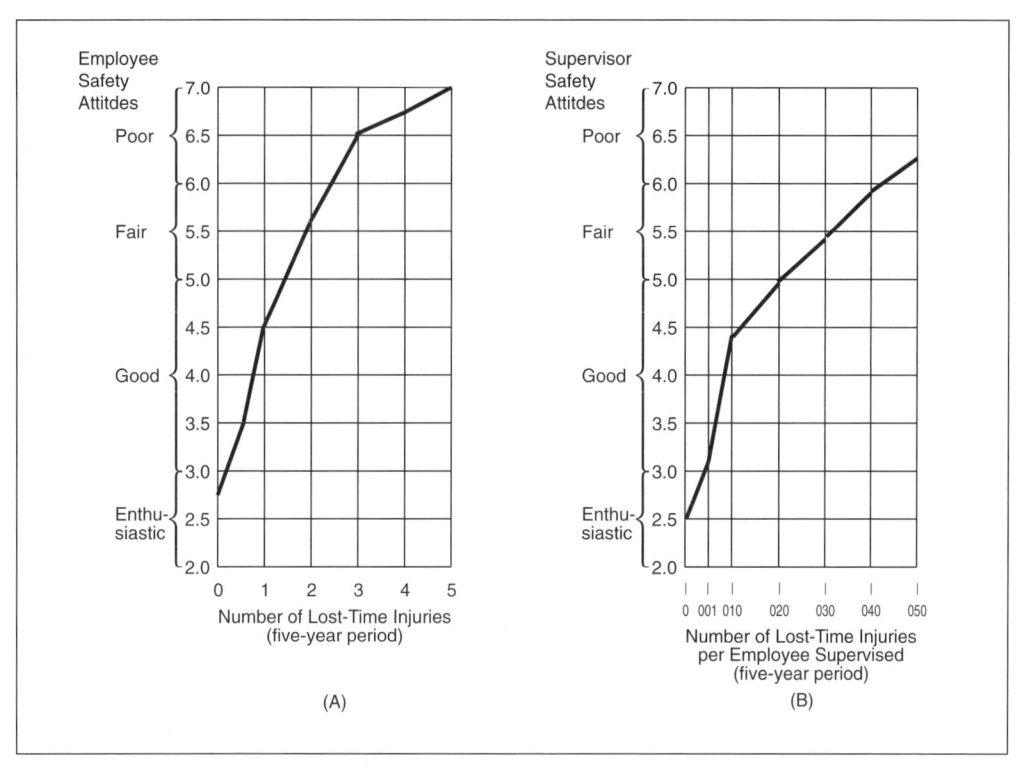

Exhibit 8-7. Relationship of employee's and supervisor's safety attitudes.

the employee, the greater the number of lost-time accidents during the five-year period studied.

The 769 male employees studied came from 47 companies representing a cross-section of various industries—companies with excellent, average, and poor safety records. Exhibit 8-7B shows that as the supervisor's safety attitude test score worsens, the number of lost-time accidents per employee under him increases. Obviously, the main conclusion to be drawn from this study is that a positive attitude toward safety fosters safe working practices.

Attitude Development

Robert Mager's book, *Developing an Attitude Toward Learning*,[7] gives us an insight into attitude development that can be directly applied to developing safety attitudes. He reports the findings of a study he made several years ago to determine students' attitudes toward different academic subjects and what formed the attitudes. According to Mager's study, a subject area tends to be favored because the person seems to do well at it; because the subject was associated with liked or admired friends, relatives, or instructors; and because the person was relatively comfortable when dealing with the subject. Conversely, a least-favored subject seems to become that because of a low aptitude for it, because it is associated with disliked individuals, and because the subject matter is associated with unpleasant conditions.

Thus, the main factors that help mold an attitude toward a subject are: the conditions that surround it, the consequences of coming into contact with it, the way that others react toward it (modeling).

Conditions

When a person receives instruction in significant subject matter (safety, for instance) he should be in the presence of as many positive and as few negative conditions as possible. If a subject that initially has no special significance is presented to someone on several occasions while he is undergoing unpleasant experiences, that subject may become a signal to escape, to get away from the unpleasantness. On the other hand, if a person is introduced to a subject while in the presence of pleasant conditions, that subject may become a signal to stick around because the person likes the association.

Mager illustrates this concept by describing the reaction of most people when a doctor moves a hypodermic needle toward them. They tend to back away or turn their heads to avoid seeing this signal of forthcoming pain. There is nothing bad about the sight of the hypodermic needle the first time we see one. But after experiencing pain while in the presence of the needle, the sight of it becomes a signal of coming pain. It is as though the mere sight of the needle becomes a condition to be avoided.

How can this concept be used with your workers? First of all, if a supervisor doesn't already know what employees consider to be good and bad conditions, he ought to find out immediately. Any form of punishment is obviously bad to them. Most forms of social interaction are good, or fun, to them. Competitions and game-type situations are good to most. Participation is good to almost everyone; being told what to do is bad, and so on.

Use what you learn about your workers' ideas of what are good and bad conditions to determine when and how to present safety instruction under the most favorable circumstances possible. If this is done, the workers should become more receptive to learning safety procedures.

Consequences

How a supervisor reacts to a worker's efforts to learn about and follow safe procedures is another important factor in determining the success or failure of a safety program. If you want to increase the probability that a response will be repeated, follow it immediately with a positive consequence. For example, if a worker always puts flammable waste material into the proper receptacle, commend him for it. Further, if a worker questions you about a safety regulation, answer him as clearly, thoroughly, and respectfully as you can, even if it is a dumb question.

Brushing aside the question or belittling the individual might keep him from talking with you about a much more important issue.

If you want to reduce the probability that the behavior will occur again, follow it immediately with an unpleasant (aversive) consequence. In other words, criticize those who do not follow good safety procedures.

This all seems obvious, but perhaps one point is a little confusing. In one case we seem to say punishment is bad, in the next breath we say it is good. Which is it? In the first case, we were talking about conditions. Here punishment is not recommended. A safety program built on a punishment atmosphere will not succeed. It is aversive. Seeing the other guy get punished creates an aversive atmosphere for you. Also, a punishment program means you must have safety rules tightly set to make punishment fair. This is aversive because it is "telling them what to do."

On the other hand, punishment following an unsafe act might work for the individual alone. But here the problem is whether we can be sure that he will no longer perform the act or he will merely make sure you don't catch him at it next time.

Aversives and Positives

What in a work situation is an aversive and what is a positive? Although it isn't always possible to know whether an event is positive or aversive for a given individual, some conditions and consequences are universal enough to provide us some direction.

First, aversives. Mager suggests we define an aversive as any condition or consequence that causes a person to feel smaller or makes his work seem inconsequential. Here are some common aversives, adapted from Mager, that might apply to safety and safety training:

Pain—not too applicable to training, but applicable to safety.
    —An accident is aversive.

Fear and anxiety—things that threaten various forms of unpleasantness, such as
    — Telling the worker by word or deed that nothing he can do will help him to succeed.
    — Telling the worker, "You won't understand this, but . . ."
    — Telling the worker, "It ought to be perfectly obvious that . . ."

- Threatening the exposure of ignorance by forcing the worker to do something in front of his group that embarrasses him.
- Being unpredictable about the standard of acceptable performance.

Frustration creators, such as
- Presenting information in larger units, or at a faster pace, than the worker can handle. (The more motivated the worker is, the greater the frustration when his efforts are blocked.)
- Speaking too softly to be heard easily (blocking the worker's effort to come into contact with the subject).
- Keeping secret the intent of the instruction, or the way in which performance will be evaluated.
- Teaching one set of skills and then testing for another.
- Testing for skills other than those stated in announced objectives.
- Refusing to answer questions.

Humiliation and *embarrassment,* for instance
- Publicly comparing a worker unfavorably with others.
- Laughing at a worker's efforts.
- Spotlighting a worker's weaknesses by bringing them to the attention of the group.
- Belittling a worker's attempt to approach the subject by such replies to his questions as "Stop trying to show off" or "You wouldn't understand the answer to that question."
- Repeated failure.
- Special classes for accident repeaters.

Boredom, caused by
- Presenting information in a monotone.
- Insisting the worker sit through instruction covering something he already knows.
- Providing information in steps so small that they provide no challenge or require no effort.
- Using only a single mode of presentation (no variety)

Physical discomfort, such as
- Allowing excessive noise or other distractions.
- Insisting the worker be physically passive for longer periods of time than he can tolerate.

Positives create an atmosphere that is more pleasant for everyone—the supervisor and the workers. There is less tension and a greater willingness to consider new ideas. Here are some positive conditions or consequences.

— Acknowledging responses, whether correct or incorrect, as attempts to learn and following them with accepting rather than rejecting comments ("No, you'll have to try again," rather than "How could you make such a stupid error!").
— Reinforcing or rewarding him for trying and succeeding.
— Providing instruction in steps that will allow success most of the time.
— Receiving learning responses in private rather than in public.
— Providing enough signposts so that the worker always knows where he is and where he is expected to go.
— Providing the worker with statements of your instructional objectives that he can understand when he first sees them.
— Detecting what he already knows and dropping that from his training. (Thus not boring him by teaching him what he already knows.)
— Providing feedback that is immediate and specific to his response.
— Giving the worker some choice in selecting and sequencing subject matter, thus making positive involvement possible.
— Providing him with some control over the length of the instructional session .
— Relating new information to old within the experience of the student.
— Treating the worker as a person rather than as a number.
— Making sure the worker can perform with ease, not just barely, so that confidence can be developed.
— Expressing genuine delight at seeing the worker succeed.

Modeling

Another way in which behavior is strongly influenced and attitudes are formed is through modeling (learning by imitation). The research on modeling tells us that if we want to teach workers behaviors, we must exhibit those behaviors ourselves. In other words, we must behave the way we want our employees to behave. When we teach one thing and model something else, the teaching is less effective than if we practice what we preach. Thus, if you want your subordinates to observe no smoking signs, for instance, you should not smoke in restricted areas.

Attitude Development

To summarize what is known about developing attitudes generally and adapting that knowledge to safety, we can say that attitude (and behavior) toward a subject may be influenced by the conditions associated with the subject matter, by the consequences of the subject matter contact, and by modeling. Furthermore, approach and avoidance behaviors are influenced by the things you do and say. But preaching, a procedure used regularly for safety, has seldom been very successful in influencing behavior.

THE GROUP

As discussed in a previous chapter, each employee decides for himself whether or not he will work, how hard he will work, and how safely he will work. He decides this based on his attitudes—attitudes toward himself, his environment, his boss, his company, his entire situation. He decides based on his knowledge, his skills, and his group's attitude toward the problem. All you can do is to create influences to help him decide, to extend some influences over what you cannot fully control, and to recognize and understand those influences over which you have no control.

Perhaps this sounds weak, as if you have little real power. This is not true. Even though you can only influence, some of the influences that you can bring to bear are powerful indeed. It is true in the final last second of decision making before the accident (or nonaccident) you have no power, but if you have used your influence well, you have done a great deal to determine whether or not the accident will occur.

Group Influence

Have you thought about the influence of the work group on the individual in that group who must make the daily decisions that can result in injury? How does the group attitude toward safety affect the individual's attitude? If safety is "sissy stuff" to his work group, how will he look at safety? If the group has decided hard hats are not to be worn, will he wear his?

Each employee is an individual, but he is also an integral part or a member of a group. Each manager must manage his crew as separate individuals and as a group. Just as chemical elements combine to make other substances with entirely different properties, individuals join to produce a group that has entirely different properties. We have to recognize the group's properties just as well as the individual's properties. A group has a distinct personality of its own. Each group makes its own decisions. The group sets its own work goals. These may be identical with management's goals, or they may be different. The group also sets its own safety standards, and it lives by its standards regardless of what management's standards are.

What Is a Group?

A group is a number of people who interact or communicate regularly and who see themselves as a unit distinct from others. Also, the members of a group are bound to one another in a state of mutual dependence. In other words, there is something at stake, and the group members share in or will share in that something. This interdependence may have nothing to do with the work task that the group performs and may be a part of the group itself or the relationships in the group. Thus, in a group, each member may depend on the others for the satisfaction of needs for affection, affiliation, or security.

The formation and development of a group depends on two things. One of these is a collection of individuals with needs and desires, and the second is a task. The most important is the social need to belong. Industrial studies have shown that those who work alone or with only one or two people are not as happy as those who work with a group.

Related to the social need for affiliation is the need to give and to receive affection. The affiliation need causes people to want to be with other people. The need for affection, the desire to be liked, causes them to conduct themselves in a way that will please those with whom they interact regularly. An additional characteristic of individuals who enter into groups is the desire to further their self-interest. The individual feels that his self-interest will be best served if he is acting with, or is at least associated with, other people. The idea is sometimes strictly a case of strength from unity, but it is more often a case of the individual feeling that if several people want and work for what he wants, his chances of success are better than if he were alone.

Whatever the situation, the formation of a group requires both a collection of individuals with certain needs and desires and a purpose. Neither is sufficient by itself. A collection of individuals without a common goal is still a collection of individuals, not a group.

Group Norms

A factor that often influences the safety of a person, but is not always understood, is the problem of what sociologists call group norms. Group norms are the informal laws that govern the way people in a group should behave and should not behave. Very often, when members of a group are asked what their norms are, they can't identify them, yet unconsciously their behavior is strongly influenced by the norms.

Group norms are the accepted attitudes about various things in the group situation. These include attitudes about how workers behave toward their boss, how they react to safety regulations, and how they react to production quotas. Norms "codify" their attitude about the company, about dress, about merit systems into recognized, accepted, and enforced behavioral patterns. If a member of a group takes on a pattern of behavior or expresses an attitude in violation of that commonly accepted by the group, there are ways of punishing him, of bringing him back into line.

In an industrial organization, if the norms developed within the work group are favorable to safety, the group itself will encourage and even enforce safe practices much better than suspension can. However, group norms often develop which are against our safety rules. A group of men might have an attitude that safety is for sissies. We don't really know why this thinking becomes entrenched, but we know it does. Often managements' first response to such a situation is to pass a regulation that will force the person to violate the norms of his peers and follow management. If the group is a strong one with a high degree of cohesiveness, the member will violate managements' direction rather than run the risk of being shut out from his work group. We ought to understand this phenomenon, and our objective should be to find some way to change the group norm and get this phenomenon working for safety, rather than against it.

Group Pressure

The power of group pressure has been shown many times in controlled experiments. Exhibit 8-8 illustrates a test with high school graduates, all men, with good eyesight. They were to tell whether two lines were the same length or not. Most of the pairs of lines were noticeably different in length.

When these men were alone, they were almost 100 percent correct in judging which line was longer. But then they became the unsuspecting victims of group pressure. Each sat in with a group, and all judged the lines. All the other people in the group were conspirators who had been told to call out wrong answers.[8]

The first two columns on the left show that the victims were only slightly misled when one or two others preceded them with the wrong answer. But when three or more others gave wrong answers before them, one-third of the victims gave in. They took the group's word rather than what their eyes told them—the majority effect. It is significant that a group of three others was as powerful as a group of 16 for misleading the victims.

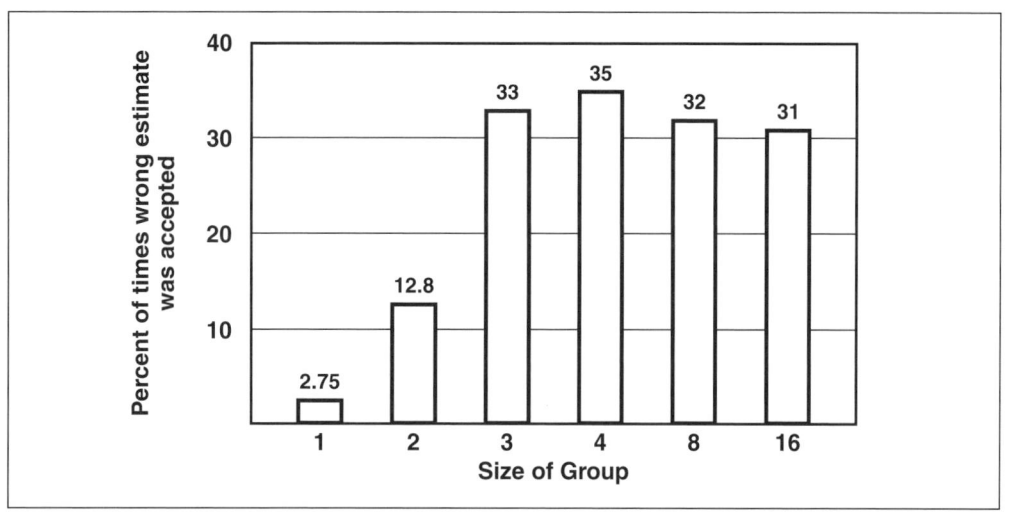

Exhibit 8-8. The size of a group in relation to its influence on individual judgment.

Some of the victims said later, after being told of the practical joke played on them, that the line lengths honestly seemed to change as they heard each of the others in the group call out the wrong answer. A few of the victims deliberately gave answers they felt were wrong; they did not want the group to think them queer. But most of the victims felt the group must be right, because it was unanimous.

Somewhat similar results occurred in tests which involved judging the size of a rectangle. Those being tested tended to change their estimates to be more like the one they were told a group of 20 to 30 people had made.

A change in one's group affiliations may cause shifts in attitudes. This was demonstrated in an analysis of 2,500 blue collar workers in a factory in the Middle West. The psychologists had records of the attitude of each of these men toward the union and toward the management. A year later the men who had been promoted to foremen or elected stewards were followed up. The men elected stewards had been no more pro-union than the rest, and the men promoted to foremen had not been any more pro-management than the others. However, within a short time after the promotion, or election, the men's attitudes had shifted. The newly elected stewards became much more pro-union than they had been formerly. The new foremen became much more pro-management than they had been a few months previously (The shifts in attitudes were larger, and more widespread, among the new foremen than among the new stewards.)[9]

Group regulations put more pressure on workers than do the standard procedures that are written in company manuals. In safety work, group pressures and group norms are perhaps the single most important determinant of worker behavior. The group sets its own safety rules, and they live by their rules, not management's.

Safety programs then must not only speak to the individual but also attempt to understand the group's safety norms and to influence those group norms to be safety oriented. Your department's safety program must help build strong work groups with goals that coincide with your safety goals.

Building Strong Groups

We can determine the strength of groups by observing some characteristic symptoms of strong and weak groups. In a strong group, the members voluntarily:

- Try to seek praise from the rest of the group
- Seek recognition from the group leaders (not management). Exert pressure on weak group members
- Put special efforts into achieving group goals

The key to identifying strong groups seems to lie in the word "voluntarily." In a strong group the members seem to want to achieve all the above. It is important to them as individuals to conform to the goals and norms of the group.

Here are some characteristic symptoms of the weak group. In such groups the members:

- Form cliques or sub-groups
- Exhibit little cooperation
- Are unfriendly
- Use no initiative
- Avoid responsibility
- Have no respect for company policies

Each manager might do well to stand back and observe his group. If he can observe any of the above identifying characteristics, perhaps he needs to try to build a stronger group. There are four elements essential to a strong group:

1. *Individual competence*—Each member of the group must have the ability to pull his own weight.
2. *Individual maturity*—Each member must be mature. This means the group dislikes the "prima donna" who can but will not do his share. They dislike the "yes man" who exerts more effort pleasing the boss than the group. They dislike the "let George do it" type.

3. *Individual strength*—Each member must have not only the ability and the maturity to do his job, he must also have the strength to earn group respect. This means no "weak sisters," no "loners."
4. *Common objectives*—Each group must have goals the members support.

In the above listing, you can see that the first three deal with qualities or characteristics of individuals that we cannot readily change. You can extend some influences of course, but basically, the strength of the group depends on the individuals you place in the group initially.

You might then ask yourself several questions before assigning individuals to groups. How did you select your employees initially? Are you looking at the competence, maturity, and strength of each person before placing him in a work group? This does not mean you cannot hire and use the loner or the prima donna. It does mean you should place him in a work situation in which he cannot destroy a strong group and not place him in a situation in which a strong group is essential to meet defined goals. What about shuffling certain individuals into other groups? At times, transfer of only a few individuals makes the difference in the strength of a group.

Your influence can be greatly felt in the development of common objectives, the fourth essential for a strong group. By doing a better job of goal setting, motivating, and communicating in regard to safety, you can make sure your goals become their chosen group goals.

Steps to Building Stronger Teams

In a classic demonstration of what forces build group cohesiveness, a group that lacked cohesiveness was transformed into a highly cooperative team within a few days. This quick building of cohesiveness was not done by asking for teamwork. No platitudes about cooperation or slogans or posters picturing the one weak link in the chain were used. The scientists just touched off the natural forces that are available when people are in groups.

The demonstration was carried out with 12-year-old boys in a secluded camp provided by the Yale University Department of Psychology. The boys did not realize they were part of an experiment. The man they thought was the caretaker of the camp was Dr. Muzafer Sherif, a psychologist. Here is an outline of the procedures by which random collections of boys were transformed into highly cohesive groups within a very short time.[10]

Physical proximity

The boys were strangers to each other at the start. After they had been in camp a few days, they were divided into two groups. Each group was in separate living quarters. This is similar to most business situations in which strangers are put together in a room separated from other work groups. Being thrown together physically presents a chance for interactions to occur that would not otherwise take place, thus providing an entering wedge for building team spirit, at least on a departmental level.

Sharing common goals

These separated groups of boys proceeded to set goals for their respective groups. They decided on decorations for and arrangement of their quarters and on other activities that appealed to their own group. Each group worked for high production on these goals and ways of reaching them. The members participated fully in deciding their shared activities. Sharing in making decisions and then working together to reach these shared goals are prime factors in building cohesiveness.

Setting up an organization and accepting leadership

These boys had not worked together for more than a few hours before they began to pool their efforts. They spontaneously organized duties within the groups.

They noticed that some members were adept at special activities, so they used these experts. They quickly divided the work and defined the responsibilities of different members. Each member soon understood what role he was expected to play. These groups also quickly came to look to a few members to play "higher roles" in coordinating the others. Captains and lieutenants emerged. These accepted leaders were from within the group not from outside it. In a business, however, the person the company designates as boss may not be the one the group would have designated. Management usually has its appointed leaders, and the group has its informal leaders.

Developing group symbols

The boys had scarcely agreed upon their accepted leaders before the members were clamoring for symbols to identify themselves as distinct groups. They invented nicknames and some jargon for their activities. Industrial groups do this if they are cohesive, and the vocabulary of one department may sound like Greek to the department down the line. The boys also developed some secrets—as offices do through the grapevines, and as families do in family jokes.

The boys' groups bought caps and T-shirts in the colors they decided on as distinctive symbols. Adults seem to have much of this same kid stuff in them. Railroaders favor a certain style of work clothes which are a trademark of their group. A house painter feels disloyal to his occupational group unless he works in painter's whites. Work clothes are part of the role that group members are expected to play.

When groups want such distinctive symbols, it is evidence of cohesiveness. But it does not necessarily follow that wearing a work uniform designed by the company will build cohesiveness.

Competing with natural enemies

The situation with boys in the experiment was such that each group quickly looked upon the other as a natural enemy. Groups tend to hold together more firmly when threatened by some enemy, or when some stress makes the members realize they are dependent on each other for security or perhaps for survival. Rivalry and stress situations are not rare within a business. One department often looks upon another as a natural enemy. One clique considers the other clique as a rival. Each clique then holds together more strongly than before, and cooperates less and feuds more with the rival clique. In case you, the appointed leader, are dogmatic and self-centered, the workers may become more cohesive, but the binding force is the goal of frustrating you rather than cooperating with you.

In the case of the boys in the experimental groups, the natural rivalry was exploited by egging the two groups into competitive contests. Little encouragement was needed; each group was itching to prove its superiority.

To intensify this rivalry, the experimenters rigged some of the contests. This made the losers furious at their opponents. Each group held closer together than ever and engaged in open as well as secret warfare. There were pitched battles in which the boy who had previously been the crybaby became the overnight hero of his group but a despised villain to the other group. To protect life and limb, it became necessary for the experimenters to order the hostilities stopped. Merely giving the order and policing the groups was not even adequate at this point.

Liking the people in the group

Many social psychologists put this item near the top of the list in building cohesiveness. You've got to like the people to become a part of the group. There is, however, a reciprocal relationship here. Cohesiveness seems to be built easiest when the people are mutually attractive to begin with. But as cohesiveness develops, people who have not previously seemed attractive become so, if they are in our group. The people in "our group" usually seem to us to be a little more capable as well as more attractive than their counterparts in competing groups.

Research and experimentation then suggest some ways to build stronger work groups. You can utilize these ways by:
— Placing people in physical proximity. If the work places do not automatically allow this (as in a shipping department) you can bring them together periodically for meetings and to participate as a group in the decision-making process.
— Allowing the group to set its own goals in safety.
— Allowing the group to organize itself in safety. Let the members select their own "enforcers," inspectors, talk-givers, committee representatives, departmental monthly safety director and so on.
— Allowing the group to develop its own symbols for itself and for safety.
— Setting up competitions with rival departments to see who can have the best safety record, inspection results, or sampling results, for instance.

DISCIPLINE[11]

Management, through its definition of policy, makes the decision that it wants safe performance from its people. Management, however does not seem to be able to force safe performance even with the most sophisticated procedures now coming on the safety scene. How can it make more effective use of discipline?

Unfortunately, the word discipline too often strikes a negative note in the minds of supervisors as well as in the minds of employees. To employees, it means rules that must be obeyed and penalties that are levied. To supervisors, it means a generally unpleasant task, the results of which almost always lead to antagonism. Yet discipline need not always be negative, it can and should be a positive process from the standpoint of both parties.

Guidelines in Using Discipline

A good supervisor tries to create a climate in which his subordinates willingly abide by company rules. But even the best supervisor cannot expect perfection; rules will still be broken. What he does about these violations will not only influence the future behavior of the employee involved, it can also have serious effects on the morale in his department and even on future contract negotiations with a union, if there is one. Here are some guidelines[11] to help you use discipline more effectively.

Know the rules and make sure your subordinates know them
You can't maintain discipline unless you know what is allowable and what is not.

Don't ignore violations
A supervisor doesn't have to issue a formal reprimand or disciplinary suspension every time a rule is broken. What he does will depend upon the nature and the circumstances of the violation and the employee's past record. The important point is that he must do something.

Get all the facts
Most arbitrated disputes are over the facts of a discipline case. As soon as a supervisor believes there has been a violation, he should establish exactly what happened.

Choose the most appropriate disciplinary action
Perhaps nothing puts a supervisor's judgment to the test more sharply than determining what discipline to give an employee who has violated a rule. He must draw the fine line between punishment that is too severe to be just and punishment that is too mild to be corrective.

Administer the discipline properly
Telling an employee he is being penalized for breaking a rule isn't any more pleasant for the supervisor than for the employee. This is the critical time for the supervisor to remember that the purpose of the discipline is corrective, not punitive.

The author of these guidelines also developed the following checklist to use when you are faced with a situation in which you must take disciplinary action.

1. **Do I have the necessary facts?**
   (a) Did the employee have an opportunity to tell his side of the story fully?
   (b) Did I check with the employee's immediate supervisor?
   (c) Did I investigate all other sources of information?
   (d) Did I hold my interviews privately to avoid embarrassing the employee?
   (e) Did I exert every possible effort to verify the information?
   (f) Have I shown any discrimination toward an individual or group?
   (g) Have I let personalities affect my decision?
2. **Have I administered the corrective measure in the proper manner?**
   (a) Did I consider whether it should be done individually or collectively?

(b) Am I prepared to explain to the employee why the action is necessary?
For instance—
—Because of the effect of the violation on the employer, fellow employees, and himself.
—To help him improve his efficiency and that of the department.
(c) Am I prepared to tell him how he can prevent a similar offense in the future?
(d) Am I prepared to deal with any resentment he might show?
(e) Have I filled out a memo for his personnel folder or a letter describing the incident, to be signed by the employee? A copy of this memo or letter should be given to the employee, and he should be told that he may respond in writing—for the record.
(f) In determining the specific penalty, have I considered the seriousness of the employee's conduct in relation to his particular job and his employment record?
(g) Have I decided on the disciplinary action as a corrective measure—not a reprisal for an offense?

3. Have I done the necessary follow-up?
(a) Has the measure had the desired effect on the employee?
(b) Have I done everything possible to overcome any resentment?
(c) Have I complimented him on his good work?
(d) Has the action had the desired effect on other employees in the department?

Punishment and Safety

Norman Maier in his text, *Psychology in Industry*, describes a study that shows the relationship between punishment and safety results.[12]

Disciplinary action is often associated with accidents or safety violations so that regardless of how a superior feels about punishment he may be involved in administering penalties. What does a supervisor do when he finds a man breaking a company regulation for which disciplinary action is specified?

Many foremen have reported to the author the dilemmas they face when a good worker commits a violation. They know that laying a man off often creates hardships, destroys friendly relations, and lowers morale. Sometimes grievances are filed, and when this occurs, their decisions are frequently reversed. They also know that they can get into trouble if they ignore violations because all guilty persons are supposed to receive the same penalty. It is not uncommon for higher supervisors to demand strict enforcement of company regulations. Some companies go so far as to have the safety department police the job because foremen are too lax. When they take this action they remove safety from the foreman's duties. Campaigns and training programs are then instituted to make foremen more safety conscious. Foremen disillusioned by such experiences resolve the dilemma by not "seeing" the violations.

Let us take a specific situation in which a foreman thinks he has found a lineman working on top of a utility pole without his safety belt. Should he attempt to determine whether a violation has occurred? If he does, he motivates the lineman to lie or to defend himself, which leads to unpleasant relations and poor cooperation. If he ignores the incident, he shirks his duty and may bring on trouble for himself with his superiors. Yet he is expected to build morale and carry out company regulations.

An experiment using 154 pairs of real foremen in a simulated situation was performed to determine what foremen do when confronted with this problem. One member of each pair acted as the foreman, the other acted as a lineman who had neglected to follow a safety regulation, but engaged his belt when he saw the foreman approaching. The foreman suspected a violation.

The results are shown in Exhibit 8-9. At point *A* the foreman decides whether or not to discuss the violation. If he decides to discuss it, he reaches point *B*. Denial of a violation will settle the matter since he cannot prove a violation. However, if the man admits the violation, the foreman moves to decision point *C* where he must decide whether or not to lay off the worker. The outcomes with regard to the objective of future safety are shown with each of the decisions in the box at the right side of the diagram.

Interestingly, 25 percent of the foremen did not discuss the violation, but instead talked generally about the company safety drive. The rest brought up the question of the violation. In these pairs, 40 percent of the linemen denied the violation and were not laid off, while 60 percent admitted it. Admission required the foreman to make a further decision. Of these, 85 percent decided not to lay off the worker as required by the company regulation, while 15 percent laid him off. From this it is very evident that various foremen in the same company see their obligations quite differently when confronted with the same situation.

Even the failure to lay off the guilty worker can get the foreman into trouble because this represents discriminatory treatment, which the union will use in cases of layoffs made by other foremen. Nevertheless, the

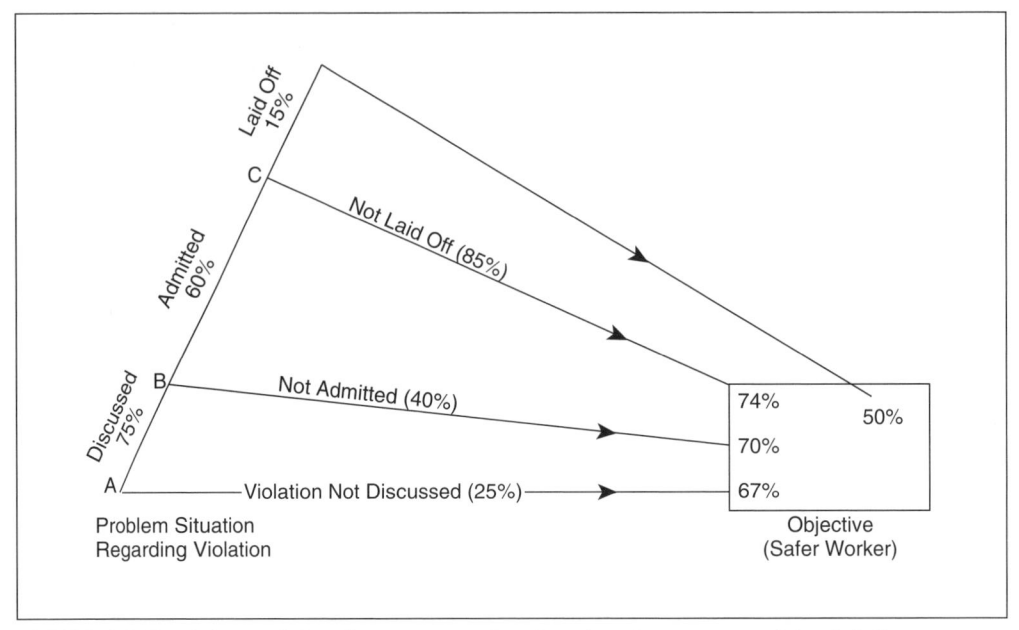

Exhibit 8-9. Decision points a foreman meets when he observes a safety violation.

majority of the foremen did not comply with the regulation in this specific situation. Although a simulated situation may not be representative of real life, the foremen regarded the results as realistic. (Simulated situations are needed to obtain statistical data on the same situation. Unfortunately real-life instances do not lend themselves to repetitions of the same situation.)

After the foremen had completed their talks with the linemen, the linemen were asked what effect this contact would have on their use of the safety belt in the future. The percentages of times that the safe practice goal was later accomplished as a result of the various decisions was very much the same (67 percent to 74 percent) for three of the decisions, but in the case of the layoff it dropped to 50 percent. Thus, as far as safety is concerned, the decision not to discuss the specific violation was the simplest way to obtain satisfactory results.

This experiment has been repeated many times. Punishment invariably produces the lowest future safe practice response. In addition, the linemen report that their production will suffer and that they will try to avoid being caught in the future. Subsequent use of this case indicates that foremen are less punitive than higher management personnel and that staff workers (including safety engineers) are the most punitive. It is one thing to have the task of administering punishment and facing the punished worker, but quite a different thing to have the impersonal task of setting up the regulations. Punishment takes on a different meaning in a general context or abstract sense than when applied to a particular individual.

Decision	Frequency Percent	Result Grievance, Percent
Full three-day layoff	34.9	45
Reduced layoff	4.6	
Warning	22.7	
Forgiven	7.5	2
Consult higher management	8.7	
Consult workers	3.5	
Other	4.6	
No decision	13.4	43

Exhibit 8-10. Decision regarding punishment for violations challenged by the union steward. *(From L. E. Danielson and N. R. F. Maier, "An Evaluation of Two Approaches to Discipline in Industry,"* Journal of Applied Psychology, *vol. 40(1956): 319-329.)*

In the foregoing experiment, the violation was in doubt. The foremen used this doubt as an excuse for not laying off the worker. They also said that the punishment (three-week layoff) was too strict.

In another simulated experiment dealing with a violation of a no-smoking regulation, the violation was made clear cut, the penalty was only a three-day layoff, and the foreman already had laid the man off. What would he do when the union steward attempted to get him to change his decision?

The results are shown in Exhibit 8-10. Only 34.9 percent of the foremen failed to alter their decision, with another 13.4 percent unable to settle the matter in the allotted time. Both outcomes resulted in a good percentage of grievances. Foremen who take a less rigid stand and do some problem-solving retain the union steward's good will. In these instances grievances are rare.

The legalistic stand of determining guilt and involving punishment, even if accepted by the company philosophy, is not supported by foremen in general. They are inclined to recognize the feelings of the workers and to think in terms of future safety and the effects on morale.

The training of foremen should be in terms of positive motivation. Obviously men do not want to have accidents, so attempts to make accidents less attractive by punishment miss the point. If people behave in unsafe ways, it is due to the presence of conflicting goals. Punishment is frequently seen as the price for getting caught and therefore often motivates men to find ways to avoid detection or to reduce the pain of the penalty. The Federal Aviation Agency, which can ground pilots for violating safety regulations, discovered that the pilots' union pays the grounded pilot his salary during the layoff, thus neutralizing the punishment.

One cannot conclude that all forms of punishment can or should be abolished. Instead, the need is to find better ways to accomplish objectives, by studying the motivational alternatives.

THE PROBLEM WORKER

Motivation, positive group attitudes, and disciplinary action can combine to produce an effective safety program. But another factor may be present that can spoil even the best record and program. This factor is the problem worker—the person who is likely to have an accident because of physical or psychological reasons.

Can We Predict?

Reducing accidents by selection and placement assumes we can foretell who will have accidents. It is fairly predictable that certain situations will be likely to produce severe injuries. However, knowing who will be injured or have accidents is quite a different thing, for this assumes that those who have accidents are in some identifiable way different from those who do not have accidents. We can examine what such identifiable ways might be by looking briefly at the various accident causation theories and at some research findings on the relationship between accidents and personal measurable factors.

We will start by looking at the terms we are using. First, there is a difference between an accident repeater and a person who is accident prone. An accident repeater is an individual who has more than one accident of the same type within a short period of time. A person who is accident prone has significantly more accidents than others. Research indicates that there is no such thing as one type of accident-prone person. Rather each individual behaves in safe and unsafe ways, depending on many things, including the environmental hazards to which he is exposed.

Current thinking in the United States is that accident proneness does exist in some people for short periods of time and in others for relatively long periods of time. In both instances, it is predictable if properly measured at the right time. If an individual has one or more accidents, it does not mean he is accident prone. Accident proneness refers to relatively consistent characteristics that make the person more susceptible to accidents. There are such people, but their number is small and their contribution to the total accident problem is slight.

The more recent studies have tended to de-emphasize the concept of accident proneness as a major cause of accidents. A survey of 27,000 industrial and 8,000 non-industrial accidents indicated that the accident repeater contributed only 0.5 percent of them, whereas 75 percent resulted from relatively infrequent experiences of a large number of persons.

Through an analysis of this survey, Dr. Morris Schulzinger came to these conclusions about accidents:

> The tendency to have accidents is a phenomenon that passes with age, decreasing steadily after reaching a peak at the age of 21. The accident rate at the age of 20 to 24, in both industrial and non-industrial areas, is two and a half times higher than at the age of 40 to 44, four times higher than at the age of 50 to 54, and nine times higher than at the age of 60 to 65.
>
> Most accidents involve young people. Seventy percent of the non-industrial accidents happen to people under the age of 35 and nearly 50 percent to those under the age of 24.
>
> Men are significantly more likely to have accidents than women. The ratio of male-to-female accidents is two to one in the non-industrial setting and, apparently, even higher in the industrial setting.
>
> Most accidents (74 percent) are relatively infrequent, solitary, experiences for large numbers of individuals (86 percent of those studied). These figures were identical in the industrial and non-industrial setting, and remained constant nearly every year for a 20-year period.
>
> Those who suffer injuries each year, over a period of three years (3 to 5 percent of the group surveyed), account for a relatively small percentage of all the accidents (0.5 percent).
>
> Irresponsible and maladjusted individuals are significantly more apt to have accidents than responsible and normally adjusted individuals.[13]

Schulzinger's studies indicated that, when the period of observation is sufficiently long, most accidents happen to individuals with a low degree of proneness. Furthermore, the relatively small percentage of the population that contributes a disproportionate number of accidents is essentially a shifting group, with new persons falling in and out of the group.

His experience suggests that in the course of a life span almost any normal individual under emotional strain or conflict may become temporarily accident prone and suffer a series of accidents in fairly rapid succession. Most persons, however, find solutions to their problems, develop defenses against their emotional conflicts, and drop out of the highly accident-prone group after a few hours, days, weeks, or months. However, some persons do remain highly accident prone throughout life, with or without lapses of years of freedom from the accident habit. The latter are the only truly accident-prone individuals. However, they contribute to only a relatively small percentage of all accidents.

Thus, the concept of accident proneness, which at first glance would seem to lead us to an improved safety record through employee selection, upon close examination seems to pose some real difficulties. We could, perhaps, screen out the tiny percent of irresponsible and maladjusted individuals who are truly accident prone, but the cost would undoubtedly be not worth it. We could hire only older workers or only females, but these approaches are, obviously, not feasible either. Or having successfully identified this year's crop of accident repeaters, we'd find that they are not next year's crop.

Physical or Mental Problems

Both physical and mental problems are often causal factors in industrial accidents and injuries. You should know how to handle these kinds of situations. Although you are not a physician or a psychiatrist, there are a few steps you can take.

> Become aware of any existing problems in your people. You can do this by merely observing them. Any distinct behavior change is usually a tip-off to a problem.
>
> Ask questions. Ask at least enough so that you know what's going on. The information will be helpful in case of a sudden serious problem and will assist medical personnel. Also, it demonstrates to the worker that you care.
>
> Seek professional medical help. Don't get in over your head—don't attempt to diagnose.

Obviously, in some situations you'll have to adjust tasks, schedules, and other aspects of the work in order to assign your people in the light of their physical and mental capabilities. If you suspect a problem, physical or mental, and if that problem is affecting a person's work, you may have to probe further and get help. Your normal observations as you supervise as well as the special techniques of inspection, observation, and interviewing will all be valuable aids in spotting potential problems.

Drug Problems

While we don't have accurate statistics on the drug problem in industry today, we do know it exists and that it is a major problem. Some studies have concluded that 6 to 8 percent of all new job applicants are drug users. Experts suggest looking for these symptoms in workers:

- Odor of burned grass on the clothing or stained fingers (from smoking marijuana)
- Watery eyes
- Musty odor, excessive use of candy and soft drinks, no appetite, upset stomach often (from using heroin)
- Long medical records
- Frequent unexplained absences
- Frequent late arrivals and early departures
- Often short of money
- Poor quality of work (many mistakes and errors)
- Increased accidents
- Sudden riches (from pushing)
- Frequent rest breaks to rest areas (for a fix)
- Possession of suspicious paraphernalia
- Sleeves rolled down in hot weather
- Dark glasses worn constantly

Your early recognition of a drug problem is the key to a successful prevention and rehabilitation program at your company. If your observation indicates a possible problem, begin to look closely at the worker's performance. Poor job performance is a reason for taking helpful action. The kinds of actions you might take are:

— Document the unsatisfactory performance—tardiness, absenteeism, errors, accidents.

— Discuss these deficiencies with the employee but discuss only the performance deficiencies on the job.

— Discuss the situation with your boss if you do not get improvement as a result of your discussions with the employee.

— Suggest the employee get medical help if you get no progress from your discussions and you are convinced it is in fact a medical problem. Do not accuse him of anything. (Obviously, if the person is incapacitated on the job, escort him to medical help.)

— Suggest that disciplinary action will be in order if improvement does not come and if the employee does not get help. If the worker's or anyone else's safety is in jeopardy, don't wait to make needed changes.

Alcohol Problems

The National Council on Alcoholism in 1968 estimated that 5.3 percent of the population were victims of an addiction to alcohol. Many of the symptoms of alcoholism are similar to those which indicate drug usage. If you note these problems, the worker may be in need of your help. Here again, you, the supervisor, are the key to an alcoholism help program in your company. You are in the best position to know what's going on.

The kinds of actions you might take with an alcohol problem are basically the same as with a drug problem. Document, discuss with the worker, discuss with your boss, suggest medical help if you don't get performance improvement, and suggest that disciplinary action might be forthcoming if improvement does not occur. Again, involve yourself only in matters relating to the employee's lack of performance on the job.

Temporary Problems

As we indicated earlier in discussing accident proneness, everybody is prone to have accidents at some time or other. There is a lot of research now on theories that will shed some light on this. One theory states that a person is much more likely to incur an injury if the person has experienced a difficult change in his life recently. Researchers have quantified the kinds of changes that might happen to a person and come up with a "danger range." If, for instance, a person has recently experienced a divorce, the death of a loved one, or a job change, the individual is much more likely to have an accident in the near future than someone who has not undergone such an experience.

Other theories suggest that, with proper observation, you can identify when an employee is in a personal crisis. During this period a person is much more likely to have an accident. Crises intervention techniques are being used in some companies to help workers through difficult periods.

Your company may be utilizing some of these newer approaches. If not, you can do much the same thing by being aware of your people's situations. Keep looking for abrupt behavior changes. These invariably mean that something is wrong somewhere.

Stress

Just as body parts wear out when subjected to overuse, people also can become ill when subjected to daily stress on the job. Psychological stress causes physical illness of all kinds (ulcers, hypertension, epilepsy, hives, and many others). If job connected, these stress-caused illnesses will cost your organization a great deal of money.

You can assist your people tremendously by paying attention to them and putting your focus on stressors they are facing on the job. You can also assist them by staying alert to whether or not they are exhibiting any of the tell-tale warning signs that show they are heading toward stress related illnesses. In the Appendix are a number of tests that you can use for yourself or your employees to identify where you (they) are in terms of coping with stress.

Notes

1. By Raymond F. Valentine, Head Performance Review Section, Management Engineering Division, Aviation Supply Office, U.S. Navy. Adapted from his article, "Problem Solving Doesn't Have to Be a Problem," *Supervisory Management* (March 1965); 23-26.
2. Chris Argyris, *Personality and Organization* (New York: Harper & Row, 1957).
3. From *New Patterns of Management* by Rensis Likert. Copyright 1961 by McGraw-Hill. All exhibits used with permission of McGraw-Hill Book Company.
4. Douglas McGregor, *The Human Side of Enterprise* (New York: McGraw-Hill, 1960).
5. Frederick Herzberg, *Work and the Nature of Man* (Cleveland: World, 1966).
6. Earle Hannaford, *Supervisor's Guide to Human Relations* (Chicago: National Safety Council, 1967), used with permission.
7. Robert Mager, *Developing and Attitude Toward Learning* (Belmont, Calif.: Fearon Publishers, 1968).
8. Solomon E. Asch, *Social Psychology* (Englewood Cliffs, N.J.: Prentice-Hall, 1952).
9. Joe Kelly, *Organizational Behavior* (Homewood, Ill.: Richard D. Irwin, Inc., 1969).
10. Muzafer Sherif, "Experiments in Group Conflict and Cooperation," *Scientific American*, Vol. 195 (Nov. 1956): 54-58.

11. Based on ideas from Walter E. Baer, "Discipline: When an Employee Breaks the Rules," *Supervisory Management* (February 1966): 20-23.
12. Norman R. F. Maier: *Psychology in Industry*, 3rd edition. Copyright © Houghton Mifflin Company, 1965. Reprinted by permission of the publishers.
13. Morris Schulzinger, *Accident Syndrome* (Springfield, Ill.: Charles C. Thomas, 1956).

CHAPTER 9 Safety Techniques That Work

Investigating for Accident Cause

SAFETY SAMPLING (SS)

SAFETY SAMPLING (SS) is a well-tested technique in accident prevention. It is a little different from the other techniques described in this section in that it normally is implemented on a company-wide basis by management or the safety director rather than within one department by the supervisor. Nevertheless, because it has been so effective in safety programs, we believe a supervisor can use this tool advantageously within his unit.

What It Is

Safety Sampling measures the effectiveness of the line manager's safety activities but not in terms of how many or how few accidents occur in his area. It measures his effectiveness before the fact of the accident by taking a periodic reading of how safely the employees are working.

Like all good accountability systems or measurement tools, Safety Sampling is also an excellent motivational tool. Each employee finds it important to be working as safely as possible when the sample is taken. Many organizations that have conducted Safety Samplings report a good improvement in their safety record as a result of the increased interest in safety on the part of employees.

Safety Sampling is based on the quality control principle of random sampling inspection, which is widely used to determine quality of production output without making 100 percent inspections. The degree of accuracy desired dictates the number of random items selected that must be carefully inspected. The greater the number inspected, the greater the accuracy.

Safety Sampling Procedure

There are four steps in a Safety Sampling:
1. *Prepare a code*—The element code list of unsafe practices is the key to Safety Sampling. This list contains specific unsafe acts that occur in your department. These are the "accidents about to happen." The element code list is developed from the past accident record and from known possible causes. The code is then placed on an observation form (see Exhibit 9-1).
2. *Take the sample*—With the code list attached to a clipboard and with a counter, start sampling. Proceed through your area, observe every employee who is engaged in some form of activity, and record whether the employee is working in a safe or an unsafe manner. Each employee is observed only long enough to make a determination. Once the observation is recorded, it should not be changed. If the observation of the employee indicates that he is performing a task safely, he is recorded on the counter. If the employee is observed performing an unsafe practice, a check is made on the form in the column that indicates the type of unsafe practice.
3. *Validate the sample*—The number of observations required to validate the sample is based on the degree of accuracy desired. Count the total number of observations you made. Determine how many unsafe practices you saw. The percentage of unsafe observations is then calculated. Using this percentage and the desired accuracy, we can calculate the number of observations required by using the data in Exhibit 9-2.
4. *Prepare the report*—The results of your observations can be presented in many different forms. However, the report should include the total percentage of unsafe activities and the number and type of unsafe practices observed.

File the reports and periodically compare the findings to spot trends in the types of unsafe practices occurring in your unit. Such information will help you plan more effective safety training programs.

Benefits

Research has proven that Safety Sampling seems to show the same trends as claim costs, number of accidents, and accident cost per manhour,

SAFETY SAMPLING (SS)

SAFETY SAMPLING		
A. Number of Safe Observations:		
B. Number of Unsafe Observations:		
Unsafe Act Noted		Sample Date
No safety glasses worn		
Improper use of tools		
Working on unguarded machine		
Not using pushsticks		
Working near tripping hazard		
Improper use of air nozzles		
Using machine improperly		
Wearing loose clothing		
Wearing rings		
Improper lifting		
Improper positioning for lift		
Climbing on racks		
Unsafe loading/piling		
Using defective equipment		
Other: (specify)		

C. Percentage

Sample Date	%	Sample Date	%	Sample Date	%

Supervisor: _____ Department: _____

Exhibit 9-1. Worksheet for sampling.

although it correlates better with the all-accident rate (all reported accidents per 1,000 man-hours).

This seems to mean that Safety Sampling provides an excellent indicator of accident problem areas before the accidents occur. Of course, by far the best value of sampling is motivational. Sampling arouses extreme interest in safety where there has been little interest before.

Percentage of Unsafe Observations	Observations Needed	Percentage of Unsafe Observations	Observations Needed
10	3,600	30	935
11	3,240	31	890
12	2,930	32	850
13	2,680	33	810
14	2,460	34	775
15	2,270	35	745
16	2,100	36	710
17	1,950	37	680
18	1,830	38	655
19	1,710	39	625
20	1,600	40	600
21	1,510	41	575
22	1,420	42	550
23	1,340	43	530
24	1,270	44	510
25	1,200	45	490
26	1,140	46	470
27	1,080	47	450
28	1,030	48	425
29	980	50	400

Exhibit 9-2. Number of observations needed for 90% accuracy. *From:* Analyzing Safety System Effectiveness, *3rd ed.* by Dan Petersen. Copyright © 1996 by Van Nostrand Reinhold.

SS Package

Reproduce Exhibit 9-1 as needed. Then:
1. Develop a code of unsafe acts you expect to find most often in your department. Record these below the common ones already listed on the form in Exhibit 9-1.
2. For the beginning of the SS program, make trips often throughout the department to build up the number of total observations. Make two or three trips a day at the start. Be sure each employee is observed on each trip.
3. Use Exhibit 9-2 to determine the total number of observations needed to ensure statistical validity. The better your people are, the more observations you'll need.
4. If the numbers in Exhibit 9-2 are too high for your department, you may have to cut them down. All this does is reduce your percentage of accuracy. This is not serious, since the results are only for your information anyway.
5. Keep track of your findings from week to week. Note progress.
6. Note any trends in types of unsafe acts that are most common. Gear your training to the most common types.

How to Set Objectives in SS

If Safety Sampling is chosen as your safety strategy, objective setting can be:
- The number of observations you will make
- The percentage of unsafe behaviors you will observe
- The frequency of samples you will make, etc.

Measurement of Performance

The measurement will be percentage of completion to goal in any of the above measurements that were chosen.

Reward

The above measures would be an integral part of the performance appraisal system, daily numbers game, etc.

STATISTICAL SAFETY CONTROL (SSC)

Statistical Safety Control (SSC) is the utilization of some common Statistical Process Control concepts in safety. The SPC tools that can be used most easily by supervisors are the Pareto Chart, the Fishbone Diagram, the Flow Chart, and the Control Chart.

SPC is a popular concept in the U.S. today. It hasn't always been. U.S. industry basically spurned the whole idea of SPC in the 1950s. Today, it wholeheartedly endorses it.

Is SPC a powerful tool? The U.S. ignored it and found their productivity dropped 15% between 1960 and 1980. The Japanese bought it and increased their productivity by 150% in the same period. And their SPC thrusts were for quality, not productivity.

In the 1980s, American managers rushed to Japan to find their secret—only to find that their secret was what the U.S. had rejected twenty years earlier.

The use of statistical methods is not an all-purpose remedy for every corporate problem. But it is a rational, logical, and organized way to create a system that can assure continuing, ongoing improvements in a quality and productivity simultaneously. This is also true for safety.

The SPC method is really a two-pronged approach—in increasing employee involvement—using some SPC tools to solve problems.

The chart in Exhibit 9-3 was developed by a group of hourly workers to identify why (combinations of reasons) they get soft tissue injuries. The problem was the soft tissue injury in wrists, shoulders, and backs.

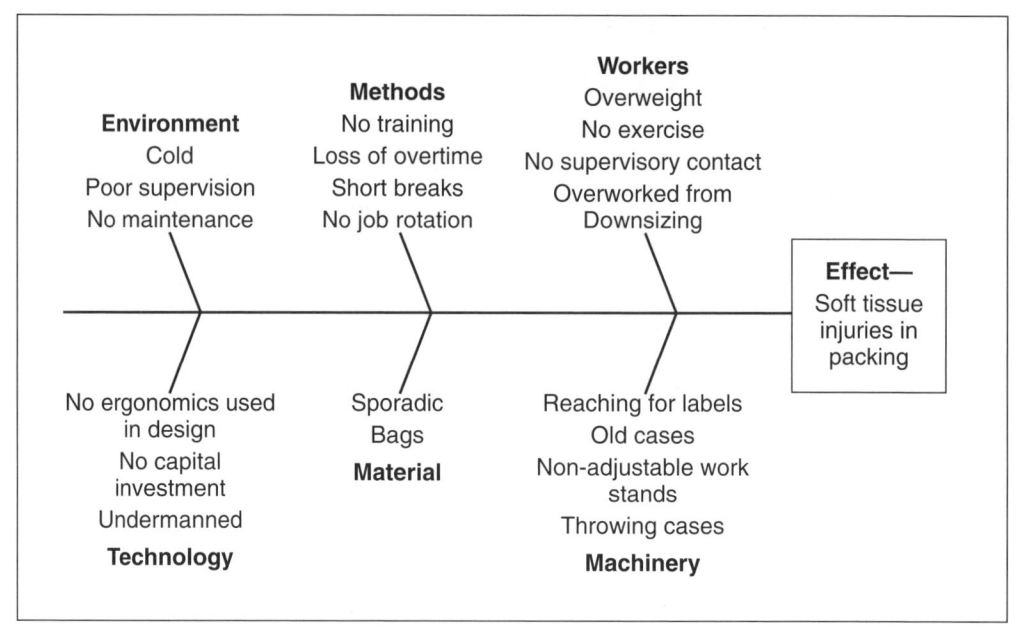

Exhibit 9-3. Fishbone chart of conditions contributing to soft tissue injuries in a packing/shipping department.

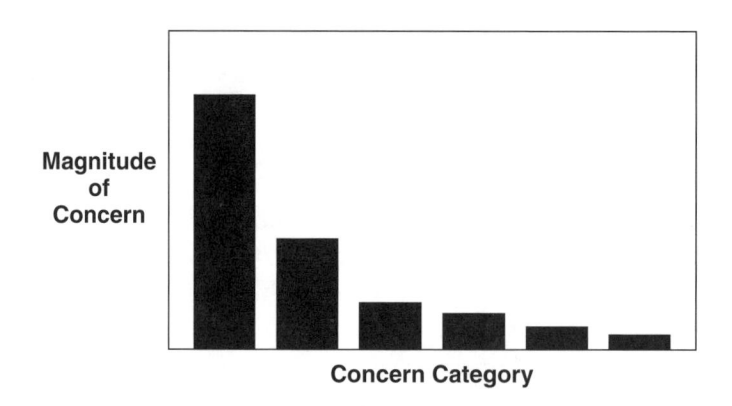

```
                    Methods          Workers
                                     Overweight
     Environment    No training      No exercise
     Cold           Loss of overtime No supervisory contact
     Poor supervision Short breaks   Overworked from
     No maintenance  No job rotation  Downsizing
                                                        ┌──────────┐
                                                        │ Effect—  │
                                                        │Soft tissue│
     ─────────────────────────────────────────────────── │injuries in│
                                                        │ packing  │
                                                        └──────────┘
     No ergonomics used  Sporadic    Reaching for labels
       in design         Bags        Old cases
     No capital          Material    Non-adjustable work
       investment                      stands
     Undermanned                     Throwing cases
     Technology                      Machinery
```

Exhibit 9-4. Basic Pareto chart: A bar graph of identified causes shown in descending order of magnitude or frequency.

Pareto Chart

Paretos can be used in many ways in safety in plotting types of injuries that have occurred in your department (see Exhibit 9-4).

From Pareto charts it was decided that a step change could be made through reducing the soft tissue injuries in two departments: the packers in the packaging department and the box handlers in shipping.

STATISTICAL SAFETY CONTROL (SSC)

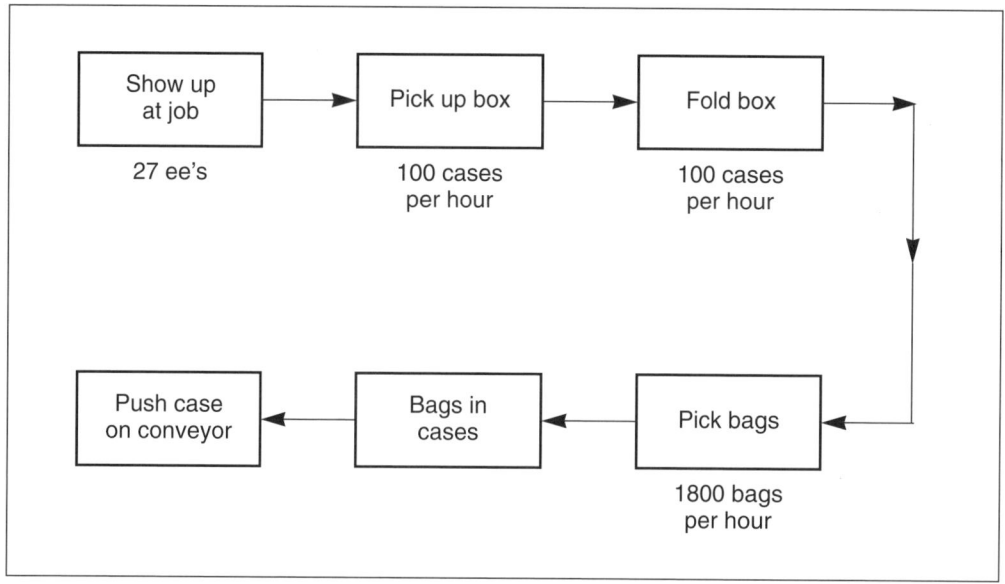

```
[Show up at job]  →  [Pick up box]  →  [Fold box]  →
   27 ee's          100 cases         100 cases
                    per hour          per hour

[Push case    ←  [Bags in    ←  [Pick bags]  ←
 on conveyor]     cases]
                                 1800 bags
                                 per hour
```

Exhibit 9-5. Flow chart of a packing process.

The group next used a third SPC tools, a Flow Chart—basically a picture of the process, step by step (Exhibit 9-5).

Finally, the group decided which factors to concentrate on:

- Get adjustable packing tables.
- Do pre-shift warm up exercises.
- Use wrist bands.
- Rotate packers hourly.

A substantial improvement was achieved. The approach became a normal way of life in the plant and was used on other safety, quality, and productivity problems.

In using SPC in safety, specific problems can be attached and solved and the system can be monitored.

How to Set Objectives in SSC

Objectives can be set in using SSC in terms of number of times the specific tools have been used, whether Pareto charts, fishbone diagrams, etc. More sophisticated measures might be percentage of accidents where fishbone diagrams were used; number of problems solved by employee groups using Paretos, flow charts, etc.

Measurement of Performance

Completion to goals set.

Reward

In the Performance Appraisal System and/or in the daily numbers game.

TECHNIQUE OF OPERATIONS REVIEW (TOR)

The Technique of Operations Review (TOR) was devised to assist you in finding some of the multiple, interrelated causes behind the accidents you are required to investigate. It is basically a tracing system. It also can be used as a training technique in safety.

In a tracing system, the investigator (you) is initially asked to identify what appears to be a major cause or factor behind an event (the accident). You select this cause from a long list of possible causes. Following the description of each major cause are numbers that lead you to other connecting factors, which might also be contributing causes.

THE TOR SYSTEM

One of the best of the tracing systems is the TOR, devised in 1967 by D. A. Weaver, while director of education at Employers Insurance of Wausau. The TOR centers around the Cause Code shown in Exhibit 9-6. The supervisor's incident investigation report in Exhibit 9-7 was developed for use with the TOR by Paul Mueller of the Green Giant Company.

The TOR begins with an incident. Its purpose is to expose what the real problems are behind this incident, which is viewed as a symptom of a more serious trouble. These are the steps of the TOR:

1. Describe the incident. State clearly what happened.
2. Select one number from the Cause Code (Exhibit 9-6) which seems to be the immediate cause of the incident.
3. Trace and eliminate. The initially chosen number is jotted on the form, and the trace step begins. Following this initial number and its description will be other numbers. Jot them down and read their descriptions. Then decide whether or not they could also be contributing causes to the incident. These numbers lead to additional numbers. List them. Read the descriptions of these and decide again. Keep tracing and eliminating numbers until they run out. When the "outs" overtake the "ins" or the final number on the list repeats the number at the top, you have come full circle. The numbers and descriptions that remain are those you have decided are contributing factors.
4. List the contributing factors.
5. Select solutions.

1 TRAINING

10 Training not formulated or need not forseen.........................23, 48, 64

11 Instruction was given but results show it didn't take................44, 47, 56, 75

12 Training available but the employee was not assigned or did not attend......26, 35, 87

13 Performance not in accord with policy or procedure...................47, 55, 62

14 Failure to provide training whose need had been specified..............34, 83, 88

15 Error blamed on faulty training when in fact the error stemmed from deficiencies in management systems.........................26, 36, 52, 81

3 DECISION & DIRECTION

30 By-passing, conflicting orders, too many bosses......................33, 48, 80

31 Decision too far above the problem.........34, 83

32 Authority inadequate to cope with the situation.................................22, 82

33 Decision exceeded authority............13, 47, 86

34 Decision evaded; power to decide not exercised.............................25, 85

35 Orders or directives failed to produce desired action. Not clear, not understood, or not followed................41, 50, 52

36 Failure to investigate, and to apply the lessons of similar mishaps.........26, 43, 61

37 Hazard or problem — controls not developed......................26, 64, 66, 86

TRACE GUIDE

10	11	12	13	14	15	
20	21	22	23	24	25	26

30	31	32	33	34	35	36	37
40	41	42	43	44	45	46	47	48

2 RESPONSIBILITY

20 Duties and tasks not clear, or not accepted..........................22, 25, 40

21 Conflicting goals.....................30, 48, 83

22 Dual or overlapping responsibility....25, 30, 48, 80

23 Pressure of immediate tasks obscures full scope of responsibilities.................10, 32, 34, 87

24 Buck passing, responsibility not tied down..........................25, 48, 82

25 Job descriptions inadequate..........48, 80, 84

26 Hazard or problem — not recognized.34, 37, 48, 81

4 SUPERVISION

40 Failure to orient or coach — new worker, unusual situation, unfamiliar equipment or process, etc........23, 24

41 Information provided was insufficient to do the job right......................23, 35, 48

42 Lack of two-way communication; failure to listen.....................15, 36, 82

43 Failure to correct behavior or procedure before accident/incident occurred.........................26, 36, 61, 73

44 Failure to supervise closely until proficiency was assured.............23, 36, 48, 64

45 Honest error. Failure to act, or action turned out to be wrong..................14, 15, 20

46 Disorder or confusion in work area.....34, 54, 56, 87

47 Job practice out of step with job training.................................15, 42, 63

48 Initiative. Failure to see problems and exert an influence on them.....23, 42, 50, 85

TOR Analysis   Worksheets and materials   Available from   R. G. Hallock, PhD
Licensing arrangements   Training Consultants to Business and Industry
Management and Employee Training   Business and Technology Center Ste. 211
                                                                                       Pueblo, Colorado 81003
                                                                                       Phone (719) 584-2435

Exhibit 9-6 (part 1). TOR Cause Code.

5 WORK GROUPS

50 Morale. Conflict, insecurity. Lack of faith in the leadership or the future of the job. 31, 56, 66, 86

51 Lax leadership in example and attitude. 31, 42, 74, 86

52 Team spirit. Failure to pull together, uncooperative. 15, 21, 74, 81

53 Rules. Not publicized, not clear. Unfair enforcement or weak discipline. ... 40, 41, 80

54 Clutter. Anything not needed in the work area. 20, 32, 60

55 Lack of things needed — tools, space, protective equipment, storage bins, etc. 24, 36, 65

56 Work group sees little advantage to themselves in doing it right 21, 82, 84

| 50 | 51 | 52 | 53 | 54 | 55 | 56 |
| 60 | 61 | 62 | 63 | 64 | 65 | 66 |

6 CONTROL

60 Work flow. Inefficient or hazardous. Layout, scheduling, stacking, piling, routing, storing. 32, 46, 48, 66

61 Unsafe condition. 36, 43, 65, 86

62 Equipment. Insufficient, unavailable, deficient design, inoperative. 26, 31, 84

63 Procedure out of step with available technology; inadequate review and revision. 24, 81, 85

64 Procedure not available or not followed. 34, 80, 88

65 Deficient inspection, reporting, or maintenance. 34, 47, 88

66 Hazard or problem — controls not maintained. 26, 34, 88

7 PERSONAL TRAITS

70 Work assignment — unsuited for this particular individual. 11, 33, 45, 63

71 Poor work habits; careless of rules, tools, equipment, procedures, etc. 26, 30, 53, 85

72 Health problem. 26, 34, 80, 81

73 Inappropriate behavior or judgement. 33, 45, 72, 83

74 Undesirable peer pressures influence work performance and risk taking. 20, 56

75 Behavior not adjusted to the workplace. ... 32, 53, 85

| 70 | 71 | 72 | 73 | 74 | 75 |
| 80 | 81 | 82 | 83 | 84 | 85 | 86 | 87 | 88 |

8 MANAGEMENT

80 Failure to assert policy and a management will before the mishap at hand. 15, 35

81 Goals are not clear or are not converted into decisions or directives. 21, 35, 85

82 Excessive span of control. 30, 36, 85

83 Conflicting priorities not resolved. Excessive emphasis on short range accomplishments. 12, 21

84 Departments inadvertently create problems for each other; inadequate coordination. 34, 36

85 Inadequate development of subordinates; failure to encourage subordinates to exercise their power to decide. 31, 50, 82

86 Inadequate appraisal and measurement of key goals and objectives. 24, 37, 50

87 Insufficient staff to cover necessary functions; failure to use available human resources or to cope with turnover and absenteeism. 23, 46, 62

88 Hazard or problem — not properly evaluated. 37, 55, 81

© 1989 by D.A. Weaver Safety Association
Pueblo, Colorado 81003
All rights reserved.

Exhibit 9-6 (part 2). TOR Cause Code.

THE TOR SYSTEM   151

	INCIDENT INVESTIGATION	
Employee (if involved)	Dept.	Clock No.
Incident date	Reported	

1. Describe the incident. Include location, witnesses, and circumstances surrounding incident. Try to identify the causal factors involved.

2. Subject causes to TOR analysis: state, trace, eliminate.

3. List factors for which you will initiate corrective action.

   Factor: Action:

4. List factors which require feasible corrective action by others. Circle routing to their attention.

Exhibit 9-7. Supervisor's incident investigation report.

Benefits

The TOR helps to identify and define problems by searching for root causes. Such causes, if undetected, might lead eventually to accidents and other kinds of management losses.

The TOR Package

Reproduce Exhibit 9-8 in sufficient quantity so that all participating supervisors receive 50 copies each. Then:

1. Take each accident and incident as it occurs in your department and submit it to the TOR analysis using the chart in Exhibit 9-6. Record your findings as you go on Exhibit 9-8.

TECHNIQUE OF OPERATIONS REVIEW	
Employee involved:	Department:
Incident date:	
1. Description of the incident:	
2. Your TOR Analysis: Trace and Eliminate	
3. Causes decided upon:	
4. Corrections decided upon:	

Exhibit 9-8. Worksheet for reviewing technique of operations.

    2. List the contributing factors and corrective actions for each accident and incident (see No. 3 on Exhibit 9-7).
    3. Prioritize the list of actions and set a time schedule to accomplish your goals.
    4. Submit to others (and to your boss) the actions that are beyond your direct control.
    5. File all Exhibit 9-8's for your future training uses.

INCIDENT RECALL TECHNIQUE (IRT)

    The Incident Recall Technique (IRT) is little used in safety. But it has great potential. It is helpful to the line manager and the corporate safety program because it unearths many more accident causes than most techniques and finds these causes before they result in accidents of a serious nature. Furthermore, the IRT involves your workers and demonstrates to them your sincere interest in creating a safe working environment. For these reasons we have included this technique and urge you to consider its use.

Why Accidents Aren't Reported

Confidential attitude surveys conducted by consulting firms in a number of companies have revealed that it is fairly common for supervisors to hide accidents. It is even more common for employees to hide them. The reasons usually given for not reporting accidents are:

- Fear of discipline
- Concern about the accident record
- Concern about reputation
- Fear of medical treatment
- Dislike of medical personnel
- Desire not to interrupt work
- Avoidance of red tape involved
- Desire to keep personal record clear
- Concern about what others might think
- Poor understanding of the need to report

Most of these barriers to reporting accidents are eliminated by the IRT. It takes little effort on your part to gain considerably more incident information on causes through the IRT than through more conventional means.

The IRT Procedure

The basic objective of the IRT is to gain the cooperation of the employee, so that he can and will freely relate all incidents from the past that he can recall. The success of the IRT, in terms of number of incidents revealed, depends primarily on the skill you use in the interview. Here are the steps of the IRT:

1. Put the employee at ease.
2. Explain the purpose of the interview and of the IRT.
3. Give assurance that the IRT is totally confidential. Your initial success depends to a great degree on this confidentiality.
4. Point out the benefits of the IRT to everyone—to the employee, his family, his department, his company.
5. Show and explain the report form to him. (See Exhibit 9-9.)
6. Conduct the interview. Simply ask him to recall each near-miss accident that he has seen or heard about on the job. With each incident recalled, be sure to determine how many times he has seen it or heard of its happening. Jot down all pertinent information.

INCIDENT RECALL TECHNIQUE		
Person interviewed:		Date:
Supervisor:		Department:
Incidents recalled:		
Analysis of causes of recalled incidents:		
Action taken on causes:		

Exhibit 9-9. Worksheet for incident recall.

7. Ask lots of questions to fill in the gaps. Avoid interrupting him but get full information on each incident.
8. Review your understanding of the incident with him. Repeat it to make sure you've got it right.
9. Discuss the causes of the incident with him and possible remedies. Make it clear you want and need his help.
10. Thank him at the conclusion of the interview.

Benefits

The benefits are obvious. First, and foremost, you may be able to prevent accidents from occurring. Second, you have heightened employee interest in safety by asking for and receiving his involvement and participation. You have also gotten more information on current accident causes than you would have through a more conventional method. And you have demonstrated to the employee that you care, that you are sincerely interested in his safety. Furthermore, the IRT can help you check your Safety Sampling findings and give you additional input for your safety training programs.

The IRT Package

Reproduce Exhibit 9-9 in sufficient quantity so that all participating supervisors can have 100 each for a start. Then:

1. Make a list of all workers in your department and schedule one interview per week.
2. Carry out the interviews as scheduled and as discussed in this section.
3. At the end of each interview complete Exhibit 9-9.
4. Let the interviewed worker review the finished report for his information and to make any additions.
5. If he wishes, provide him with a copy of the filled-in form.
6. Keep several additional copies for your records and for future training material.

CHAPTER 10 Safety Techniques That Work

Inspecting for Hazards

JOB SAFETY ANALYSIS

JOB SAFETY ANALYSIS (JSA) is a procedure that identifies the hazards associated with each step of a job, and develops solutions for each hazard that either eliminate it or control it. A job safety analysis worksheet is illustrated in Exhibit 10-1. In the left column, the basic steps of the job are listed in the order in which they are performed. The middle column describes how to perform each job step. The right column gives the safety procedures that should be followed to guard against hazards. The basic steps in making a JSA are:

    1. Select the job to be analyzed
    2. Break the job down into successive steps
    3. Identify the hazards and potential accidents
    4. Develop ways to eliminate the hazards listed

A blank worksheet used for OSHA compliance is shown in Exhibit 10-2.

Selecting the Job

In selecting jobs to be analyzed and in establishing the order of job priorities, the National Safety Council suggests that you be guided by these factors:

— Frequency of accidents. A job that has a repeated number of accidents is a good candidate for an early JSA.

JOB:		
What to Do (*Steps in sequence*)	**How to Do It** (*Instructions*) (*Reverse hands for left-handed operator.*)	**Key Points** (*Items to be emphasized.* *Safety is always a key point.*)
1.	1.	1.
2.	2.	2.
3.	3.	3.
4.	4.	4.
5.	5.	5.
6.	6.	6.

Exhibit 10-1. Job analysis worksheet.

— Severity of accidents or injuries. Any job that has produced disabling injuries might be considered for an early JSA.

— A high potential for severity. If the potential for a serious accident or injury is present, a JSA might well be warranted.

— New jobs or changed jobs.

Breaking the Job Down

First, the job is broken down into its basic steps. These steps should describe what is being done in order of occurrence. The National Safety Council suggests these key points in breaking down a job:[1]

1. **Select the right man to observe.** The man selected must be experienced, capable, cooperative, and willing to share his thoughts.
2. **Brief him on your purpose.** If the worker selected has never worked on a JSA before, thoroughly explain it to him. Ask him for his help.

```
Job operation _____
Presently required personal protective equipment _____
Sequence of job steps              Hazards or OSHA violations
1.                                 1.
2.                                 2.
3.                                 3.
4.                                 4.
Recommended safe procedure
1.
2.
3.
4.
```

Exhibit 10-2. OSHA job safety analysis worksheet. *(From* The OSHA Compliance Manual *by Dan Petersen. Copyright © 1975 by McGraw-Hill. Used with permission of McGraw-Hill Book Company.)*

3. **Observe the job** for the breakdown,
4. **Record each step** on the worksheet,
5. **Check the breakdown with the worker** when you are done and get his input.

Identifying Hazards

After the breakdown, each step should be analyzed by you in detail to identify hazards and potential accidents. Each should be recorded on the worksheet in the center column. Keep hazards parallel with the steps recorded. Check with the employee you're working with for his ideas. Check also with other employees who have knowledge of that job.

Developing Solutions

When the hazards have been identified, your next step is to begin to develop solutions to the problems you've identified. Solutions might incorporate:

— An entirely different way to do the job;
— A change in physical conditions, layout, or environment;
— A changed job procedure;
— A change in frequency of how often the job is performed.

For each hazard on the sheet, ask "What can be done differently and how should it be done?" Answers and solutions should be very specific and

JOB SAFETY ANALYSIS				
Job:			Date:	
Employee:			Supervisor:	
Job Steps		Hazards/Potential Accidents		Controls

Exhibit 10-3. Worksheet for analyzing job safety.

very concrete to be of value. Solutions which merely state "Be more alert" or "Use More caution" or something similar are worthless. Solutions should state exactly what to do and how to do it.

While performing the JSA, you'll learn more about the job observed than ever before. You also will have involved an employee and demonstrated to him and the department that you care about his safety on that job. And you'll be creating safer conditions for the job observed.

JSA Procedure

Reproduce Exhibit 10-3 in sufficient quantity. Then set a schedule for your department, as follows:

1. **List all jobs in the department.**
2. **Schedule them for analysis by you.** Take no more than perhaps three per week. Take no less than one per week
3. **Carry out each analysis as outlined in this section.**
4. **Upon completion of each analysis, review it in detail with the worker involved.** He may have valuable additions.
5. **Provide him with a copy of the final analysis.**
6. **Keep several copies yourself** and use these in future orientation and training sessions.

How To Set Objectives With JSA'S

Number of JSA's to be completed.

Measurement

Completion of goal.

Reward

Performance Appraisal System and daily numbers game.

HAZARD HUNT (HH)

One other method that has been used successfully to spot possible accident causes is the Hazard Hunt. It is also good for involving your people. To implement the procedure follow these steps:

1. **Make copies of the Hazard Hunt form (Exhibit 10-4) available to all of your people.**
2. **Hold a short session with them to explain the form and the reasons for using it.**
3. **Have employees jot down anything they feel is a hazard and return the form to you.**
4. **Review the forms, correct those hazards you can, and initiate action when correction is out of your control.**
5. **Inform employees of your actions.** Always tell them what you are doing, even if you decide the hazards they mention are not really problems. Then hash over any disagreements with individuals to clear the air.
6. **If you agree something is a hazard and must be corrected, assign a priority to it and schedule it for rectification.** The matrix in Exhibit 10-5 provides a good guide for prioritizing.

```
To: _____
From: _____
                    HAZARD HUNT
I think the following is a hazard: _____
_____
_____

DO NOT WRITE BELOW HERE—TO BE FILLED IN BY SUPERVISOR
Supervisor:
    Agree this is a hazard.
        Corrected by Supervisor on _____ Discussed on _____
        If can't correct, sent to Personnel on _____
        Job order on _____ Scheduled _____ Discussed on _____
    Do not agree this is a hazard.
        Discussed _____ Conclusion _____
        To Personnel _____ Conclusion _____
DO NOT WRITE BELOW HERE—FOR PERSONNEL USE.
Supervisor _____ HH # _____
Matrix # _____ (Seriousness)
```

Exhibit 10-4. A Hazard Hunt form. *(From* The OSHA Compliance Manual *by Dan Petersen. Copyright © 1975 by McGraw-Hill. Used with permission of McGraw-Hill Book Company.)*

The HH Procedure

Reproduce Exhibit 10-4 in sufficient quantity so that each participating supervisor can have several hundred copies. Then:

1. **Hold a short meeting with all the workers in your department.** Explain that you are asking them for their help in finding hazards and OSHA violations. Ask them to look around their area and the whole department and jot down on the Hazard Hunt form any problems they see. Have them return the forms to you.
2. **Give five to ten copies of the form to each.**
3. **React to every form you get back—give an answer to each employee who turns in a form.** Tell him what the company will do about each hazard he listed. If nothing can be done, tell him why.

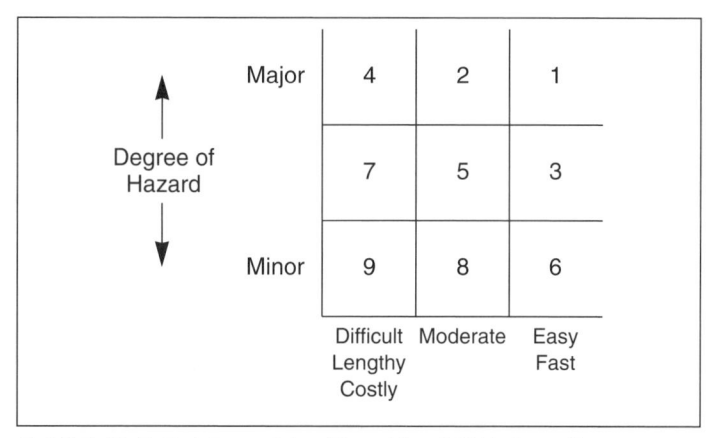

	Difficult Lengthy Costly	Moderate	Easy Fast
Major	4	2	1
	7	5	3
Minor	9	8	6

(Degree of Hazard — vertical axis)

Exhibit 10-5. Safety matrix. *(From* The OSHA Compliance Manual *by Dan Petersen. Copyright © 1975 by McGraw-Hill. Used with permission of McGraw-Hill Book Company.)*

4. Use the matrix to explain to an employee why it will take a while to correct the problem he listed, if it has a low priority.

5. Keep the forms on file for future training purposes.

How To Set Objectives With HH's

Number of HH's submitted.

Measurement

Percentage to goal.

Reward

Performance Appraisal System and daily numbers game.

OSHA COMPLIANCE CHECK

In addition to preventing accidents, the supervisor also has a responsibility to ensure his organization complies with the law. The OSHA Compliance Check does this.

OCC Procedure

The first step in the OCC is to develop a checklist of the specific OSHA standards that apply in your department. To do this you, probably will first obtain a copy of the standards and begin to sort through them to determine which will apply to you. You may be able to get staff assistance for this.

These standards are very comprehensive. To illustrate, following is the table of contents for Subparts D-S and Z of 29 CFR Part 1910—Occupational Safety and Health Standards for General Industry.

Subpart A-General

1910.1	Purpose and scope
1910.2	Definitions
1910.3	Petitions for the issuance, amendment, or repeal of a standard
1910.4	Amendments to this part
1910.5	Applicability of standards
1910.6	Incorporation by reference
1910.7	Definition and requirements for a nationally recognized testing laboratory
1910.8	OMB control numbers under the Paperwork Reduction Act

Subpart B—Adoption and Extension of Established Federal Standards

1910.11	Scope and purpose
1910.12	Construction work
1910.15	Shipyard employment
1910.16	Longshoring and marine terminals
1910.17	Effective dates
1910.18	Changes in established Federal standards
1910.19	Special provisions for air contaminants

Subpart C—General Safety and Health Provisions

[Removed and Reserved.]

Subpart D—Walking-Working Surfaces

1910.21	Definitions
1910.22	General requirements
1910.23	Guarding floor and wall openings and holes
1910.24	Fixed industrial stairs
1910.25	Portable wood ladders
1910.26	Portable metal ladders
1910.27	Fixed ladders
1910.28	Safety requirements for scaffolding
1910.29	Manually propelled mobile ladder stands and scaffolds (towers)
1910.30	Other working surfaces

Subpart E—Means of Egress

1910.35	Definitions
1910.36	General requirements
1910.37	Means of egress, general
1910.38	Employee emergency plans and fire prevention plans
	Appendix to Subpart E—Means of Egress

Subpart F—Powered Platforms, Manlifts, and Vehicle-Mounted Work Platforms

1910.66	Powered platforms for building maintenance
1910.67	Vehicle mounted elevating and rotating work platforms
1910.68	Manlifts

Subpart G—Occupational Health and Environmental Control

1910.94	Ventilation
1910.95	Occupational noise exposure
1910.97	Nonionizing radiation
1910.98	Effective dates

Subpart H—Hazardous Materials

1910.101	Compressed gases(general requirements)
1910.102	Acetylene
1910.103	Hydrogen
1910.104	Oxygen
1910.105	Nitrous oxide
1910.106	Flammable and combustible liquids
1910.107	Spray finishing using flammable and combustible materials
1910.108	Dip tanks containing flammable or combustible liquids
1910.109	Explosives and blasting agents
1910.110	Storage and handling of liquefied petroleum gases
1910.111	Storage and handling of anhydrous ammonia
1910.119	Process safety management of highly hazardous chemicals
1910.120	Hazardous waste operations and emergency response

Subpart I—Personal Protective Equipment

1910.132	General requirements
1910.133	Eye and face protection
1910.134	Respiratory protection
1910.135	Head protection
1910.136	Foot protection
1910.137	Electrical protective equipment
1910.138	Hand protection
	Appendix A to Subpart I—References for further information
	Appendix B to Subpart I—Nonmandatory guidelines for hazard assessment and personnel protective equipment selection

Subpart J—General Environmental Control

1910.141	Sanitation
1910.142	Temporary labor camps
1910.144	Safety color code for marking physical hazards
1910.145	Specifications for accident prevention signs and tags
1910.146	Permit required confined spaces
1910.147	The control of hazardous energy (lockout/tagout)

Subpart K—Medical and First Aid

1910.151	Medical services and first aid

Subpart L—Fire Protection

1910.155	Scope, application and definitions applicable to this subpart
1910.156	Fire brigades
1910.157	Portable fire extinguishers
1910.158	Standpipe and hose systems
1910.159	Automatic sprinkler systems
1910.160	Fixed extinguishing systems, general
1910.161	Fixed extinguishing systems, dry chemical
1910.162	Fixed extinguishing systems, gaseous agent
1910.163	Fixed extinguishing systems, water spray and foam
1910.164	Fire detection systems
1910.165	Employee alarm systems

Appendix A to Subpart L—Fire Protection
Appendix B to Subpart L—National Consensus Standards
Appendix C to Subpart L—Fire Protection References for Further Information
Appendix D to Subpart L—Availability of Publications Incorporated by Reference in Section 1910.156 Fire Brigades
Appendix E to Subpart L—Test Methods for Protective Clothing

Subpart M—Compressed Gas and Compressed Air Equipment

1910.169	Air receivers

Subpart N—Materials Handling and Storage

1910.176	Handling materials general
1910.177	Servicing multipiece and single piece rim wheels
1910.178	Powered industrial trucks
1910.179	Overhead and gantry cranes
1910.180	Crawler locomotive and truck cranes
1910.181	Derricks
1910.183	Helicopters
1910.184	Slings

Subpart O—Machinery and Machine Guarding

1910.211	Definitions
1910.212	General requirements for all machines
1910.213	Woodworking machinery requirements
1910.215	Abrasive wheel machinery
1910.216	Mills and calenders in the rubber and plastics industries
1910.217	Mechanical power presses
1910.218	Forging machines
1910.219	Mechanical power transmission apparatus
1910.220	Effective dates
1910.221	Sources Of standards
1910.222	Standards organizations

Subpart P—Hand and Portable Powered Tools and Other Hand-Held Equipment

1910.241	Definitions
1910.242	Hand and portable powered tools and equipment, general
1910.243	Guarding of portable powered tools

1910.244   Other portable tools and equipment

Subpart Q—Welding, Cutting and Brazing

1910.251   Definitions
1910.252   General requirements
1910.253   Oxygen fuel gas welding and cutting
1910.254   Arc welding and cutting
1910.255   Resistance welding

Subpart R—Special Industries

1910.261   Pulp, paper and paperboard mills
1910.262   Textiles
1910.263   Bakery equipment
1910.264   Laundry machinery and operations
1910.265   Sawmills
1910.266   Logging operations
1910.268   Telecommunications
1910.269   Electric power generation, transmission, and distribution
1910.272   Grain handling facilities

Subpart S—Electrical

1910.301   Introduction
1910.302   Electric utilization systems
1910.303   General requirements
1910.304   Wiring design and protection
1910.305   Wiring methods, components, and equipment for general use
1910.306   Specific purpose equipment and installations
1910.307   Hazardous (classified) locations
1910.308   Special systems
1910.331   Scope
1910.332   Training
1910.333   Selection and use of work practices
1910.334   Use of equipment
1910.335   Safeguards for personnel protection
1910.399   Definitions applicable to this subpart
            Appendix A to Subpart S—Reference documents

Subpart Z—Toxic and Hazardous Substances

1910.1000   Air contaminants
1910.1001   Asbestos
1910.1002   Coal tar pitch volatiles; interpretation of term
1910.1003   4-Nitrobiphenyl
1910.1004   alpha-Naphthylamine
1910.1006   Methyl chloromethyl ether
1910.1007   3,3-Dichlorobenzidine (and its salts)
1910.1008   bis-Chloromethyl ether
1910.1009   beta-Naphthylamine
1910.1010   Benzidine
1910.1011   4-Aminodiphenyl
1910.1012   Ethyeneimine

1910.1013 beta-Propiolactone
1910.1014 2-Acetylaminofluorene
1910.1015 4-Dimethylaminoazobenzene
1910.1016 N-Nitrosodimethylamine
1910.1017 Vinyl chloride
1910.1018 Inorganic arsenic
1910.1020 Access to employee exposure and medical records
1910.1025 Lead
1910.1027 Cadmium
1910.1028 Benzene
1910.1029 Coke oven emissions
1910.1030 Bloodborne pathogens
1910.1043 Cotton dust
1910.1044 1,2-dibromo-3-chloropropane
1910.1045 Acrylonitrile
1910.1047 Ethylene oxide
1910.1048 Formaldehyde
1910.1050 Methylenedianiline
1910.1051 l,3-Butadiene
1910.1096 Ionizing Radiation
1910.1200 Hazard communication
1910.1201 Retention of DOT markings, placards and labels
1910.1450 Occupational exposure to hazardous chemicals in laboratories

It is immediately obvious that not all of these sections apply to your department. Go through the list and identify which sections apply to you.

Obtain your copy, read the applicable sections and develop your own checklist as shown in Exhibit 10-6.

After developing the checklist, share it with your people. Ensure each week someone (you or one of your group) checks the department based on your checklist.

How to Set Objectives With OCC's

Here you may use a number of checks made, number of checks per week, whether or not the checklist is completed; number of violations found, number of violations corrected, etc.

Measurement

Completion to goal.

Reward

Performance Appraisal System and daily numbers game.

| SUBPART I—Personal Protective Equipment | INSPECTED BY _____ |
| SECTION 1910.132 General Requirements | DATE _____ |

ITEM NO. SUB PARAG.

a   (a) *Application.* Protective equipment, including personal protective equipment for eyes, face, head, and extremities, protective clothing, respiratory devices, and protective shields and barriers, shall be provided, used, and maintained in a sanitary and reliable condition wherever it is necessary by reason of hazards of processes or environment, chemical hazards, radiological hazards, or mechanical irritants encountered in a manner capable of causing injury or impairment in the function of any part of the body through absorption, inhalation or physical contact.

SECTION 1910.133 Eye and Face Protection

a   (a) *General requirements.* (1) The employer shall ensure that each affected employee uses appropriate eye or face protection when exposed to eye or face hazards from flying particles, molten metal, liquid chemicals, acids or caustic liquids, chemical gases or vapors, or potentially injurious light radiation.

Exhibit 10-6. OSHA Check Off Sheet.

ERGONOMIC ANALYSIS

An ergonomic analysis (EA) is a way of reducing the probability that your people will be injured through exposure to cumulative trauma or repetitive motion. Each job your people do should be analyzed.

This ergonomic analysis has as its primary focus the preventing of Repetitive Motion Injuries, which are commonly called Cumulative Trauma Disorders (CTD's). These CTD's are occupational injuries that develop over time, affecting the musculoskeletal and peripheral nervous systems. They can develop in any part of the body but are most prevalent in the arms and back. These injuries are caused by jobs that require repeated exertions and movements near the limits of the individual's strength and range-of-motion capability. These movements, although not initially painful, cause microtraumas to the soft tissues. Over time, small strains to the muscle/tendon/ligament system build up, resulting in fatigue and soreness. If the individual continues the action that is causing pain, cumulative trauma disorders are likely to develop. CTD's can have the following affects on individuals:

- pain
- numbness of loss of sensation
- reduced strength
- degraded ability to perform work
- degraded ability to participate in leisure activities.

Here are several good reasons why we should try to eliminate CTD's from the workplace:

- it's the right thing to do
- it's the humane thing to do
- CTD's are costly
- OSHA will make you.

In order to prevent CTD's we have to understand what causes them. Cumulative trauma disorders have been correlated with hazardous combinations of the following "generic" risk factors:

- forceful exertions
- frequent exertions
- body posture
- mechanical stress
- vibration
- low temperatures.

The presence of the above risk factors are related to the design of the job. For example, we can either design a workstation that requires the worker to use a poor posture all day or we can design it so that all operators can do the operation in a good posture. Areas of the design that may expose the worker to the generic risk factors are called "job specific" risk factors and include items on the following list:

- workplace layout
- tools
- parts
- environment

ERGONOMIC ANALYSIS

Hand & Wrist	Possible CTD Problem	Neck	Possible CTD Problem
GRASP • pinch grip • static hold	• prolonged pinch grip • forceful grasp	**NECK POSTURE** • bend/twist > 20°	• > 50% of time
WRIST POSTURE • flexion/extension • ulnar/radial deviation	• flexion/extension >45° • radial/ulnar deviation	**Back** **LIFTING**	**Possible CTD Problem** • > 90% of Action Limit • non-NIOSH
FREQUENCY • hand or wrist manipulations	• > 10 per minute	**TORSO POSTURE** • torso bending • torso twisting	• bend > 45° • bend > 20° + twist
MECHANICAL STRESS • localized pressure to palm or fingers • scraping/pumping • strike with hand • single finger trigger	• prolonged exposure • intense exposure	**STATIC HOLD/CARRY** • > 5 seconds	• > 10 lbs. • 5-10 lbs. with flexed shoulder
VIBRATION • high frequency vibration	• prolonged exposure	**STATIC LOAD** • not able to change sit/stand posture over work day	• poor posture
Arm & Shoulder	**Possible CTD Problem**	**PUSH/PULL** • whole body action	• poor conditions
ARM WORK • exertions > 5 lbs.	• little rest between exertions • large exertion	**Legs** **FOOT ACTUATION** • foot pedals	**Possible CTD Problem** • excessive force • extreme posture • high frequency or duration
STATIC LOAD • prolonged holding	• unsupported • large exertion	**LEG POSTURE** • knee • ankle	• deep squat • kneeling • 1 legged posture • walk/stand on uneven surface
ELBOW POSTURE • fully flexed • fully extended • rotated forearm	• repeatedly		
SHOULDER POSTURE • flexed • extended • abducted	• flex/abduct > 90° • any extension	**MECHANICAL STRESS** • localized pressure • kicking	• high force • prolonged exposure
MECHANICAL STRESS • sharp edges • hard surfaces	• repeatedly • high force		

Exhibit 10-7. Ergonomic Analysis overview.

Exhibit 10-7 is a table showing for each involved part of the body each of the factors that must be looked at on an ergonomic analysis, and at what point you should begin to be concerned that a factor could eventually cause a CTD.

EA Procedure

With the information in Exhibit 10-7 in mind, look at each job to determine the potential of a CTD. It may help to reproduce the form in Exhibit

Company			CHECK IF A POSSIBLE CTD PROBLEM
Dept.			
Supervisor			
Job Name			

LEFT HAND & WRIST	
Grasp	
Wrist Posture	
Frequency	
Mechanical Stress	
Vibration	

Date ....
Time ....
Pace: mach.   self
Job Rotation: yes   no
Regulator Operator yes   no
Tools/Parts   Weight

RIGHT HAND & WRIST	
Grasp	
Wrist Posture	
Frequency	
Mechanical Stress	
Vibration	

COMMENTS

ARM & SHOULDERS	
Arm Work	
Static Load	
Elbow Posture	
Shoulder Posture	
Mechanical Stress	

NECK	
Neck Posture	

BACK	
Lilfting	
Torso Posture	
Static Hold/Carry	
Static Load	
Push/Pull	

LEGS	
Foot Actuation	
Leg Posture	
Mechanical Stress	

Exhibit 10-8. Ergonomic Analysis checklist.

10-8 which is the checklist you carry with you with Exhibit 10-7 on the back for your constant reference.

How to Set Objectives with EA's

Objectives can be set with number of analyses made, percentage of jobs completed; number of suggestions made; number of changes made; number of completions, etc.

Measurement

Completion to goal.

Reward

Performance appraisals; daily numbers game.

CHAPTER 11 Safety Techniques That Work

Coaching

JOB SAFETY OBSERVATION (JSO)

A **Job Safety Observation (JSO)** provides you with a device to learn more about the work habits of each of your people. Following the procedures described below, you can use this opportunity to check on the results of past training; make immediate, on-the-spot corrections and improvements in work practices; and compliment and reinforce safe behavior. Through your comments you can encourage proper attitudes toward safety. Follow these steps when implementing a JSO program:
1. **Select the worker and job to be observed**
2. **Make the observation**
3. **Record**
4. **Review the results with the employee observed**
5. **Follow up.**

Worker Selection

Eventually you'll observe all your people. But you might consider the following possibilities when determining which person to observe first:
- The new person on the job
- People recently out of training programs
- Below-average performers
- Accident repeaters
- Risk takers
- Workers with special problems.

Making the Observation

You should in most cases tell the person to be observed what you are doing prior to the JSO. Then simply observe him in his normal operation making any applicable notes on the worksheet about his normal work practices and procedures. Be careful to stay out of his way and don't distract him.

Recording

Fill out the Job Safety Observation Worksheet during the JSO after reviewing it with the worker. File it in any manner that suits your needs.

The Review

When you have completed the JSO sit down with the worker and give him your conclusions. Express appreciation to him for cooperating and lay out your honest feelings about his work habits and practices. The first time you go through this with a worker both of you will be nervous, and he may be concerned or apprehensive. Keep him at ease and keep the discussion informal and friendly. Do not let the discussion be a one-way communication. Encourage him to talk and give you his views as well as any problems and any barriers he sees to working safely.

The Follow-up

Follow up the observation as needed. In some cases follow-up will be often. In others, it will be seldom. How often depends on the man and on the job. Follow-up JSO's are usually a good idea after a job change.

Benefits of JSO's

The JSO is a feedback device. It provides excellent information on the effectiveness of your training and on the adequacy of your job procedures. Through the JSO any substandard practices can usually be identified before an accident happens. JSO's also give you the opportunity to sit down with your people individually to discuss their performance and to compliment or correct their work habits. In addition, you get to know each worker better and thus can spot any physical or psychological problems more readily.

JSO Procedure

Reproduce Exhibit 11-1 in sufficient quantity so all supervisors involved in the JSO program will have at least 25 copies. Then set a schedule for your department as follows:

JOB SAFETY OBSERVATION			
Employee:		Supervisor:	
Job:		Date:	Time:
Notes on any job practices that are unsafe:			
Notes on any practices that need change or improvement:			
Notes on any practices that deserve complimenting:			
Notes on the review and discussion:			

Exhibit 11-1. Worksheet for observation.

1. List those activities performed in your department that you feel warrant a JSO.
2. Schedule the JSO's. Do no more than one per day. Do no less than one per week.
3. Carry out each observation as outlined in this section.
4. Upon completion of each JSO, review it with the worker observed
5. Provide the worker with a copy of the completed form.
6. Keep several copies in your files for future training purposes

How to Set Objectives with JSO's

Objectives can be set with number of JSO's made, percentage of people covered, number of suggestions made; number of positive strokes given, etc.

Measurement

Completion to goal.

Reward

Performance Appraisal System, daily numbers game, etc.

ONE-ON-ONE CONTACTS

The one-on-one contact (OO) is a personal contact between you, the supervisor, and each person that works for you. The contact can be safety related, or it can be used simply as an opportunity for you to get to know each person that works for you a little better. Contacts must be often to be useful.

Procedure

Schedule yourself time each day to make your contacts. Contact at least one person each day. Keep a list of your people to ensure you miss no one over time. If you wish, use the Worker Safety Analysis Form in Chapter 12 to assist.

How to Set Objectives with OO's

Number of OO's per day, week, etc.

Measurement

Completion to goal.

Reward

Performance Appraisal System and daily numbers game.

SAFE BEHAVIOR REINFORCEMENT

It has been known through research for many years that positive reinforcement immediately following a desired behavior is the strongest way to build and maintain safe behavior. Safe Behavior Reinforcement (SBR) is simply a way of utilizing this fact.

SBR Procedure

Each day ensure that you schedule yourself some time to observe each person that works for you at least once. When you observe, react to whatever you see. If the person is working unsafely, react as you normally would.

If, however, the person observed is working safely, make an immediate contact with that person, and say something positive—pleasurable (an "attaboy" or whatever fits your personality) attaching it to the behavior you have observed. Make sure that the person knows that you desire safe work, and will be looking for it each day.

Keep track each day of the number of observations made and the number of positive reinforcements. Total them each week, month, etc., and publish them for your group.

How to Set Objectives with SBR

Number of observations, number of positive reinforcements; percentage of positive reinforcements versus negative reinforcements to achieve desired behavior; percentage of people covered, etc.

Measurements

Completion to goal (through self reporting).

Reward

Performance Appraisal System; daily numbers game.

ONE-MINUTE SAFETY SYSTEM

The One-Minute Safety (OMS) system is a three-step process.
1. Sit down with each subordinate and agree on what are the key objectives of his/her job. Agree also as to the safety hazards, critical behaviors and your wishes.
2. Check with each employee at least once each week, and spend a minimum of one minute with each on a one minute praising.
3. Check at least once each week to see if a one minute reprimand is in order—make it positive also by using only 15 seconds to point out the behavior you want changed.

Schedule yourself to ensure each person that works for you is checked each week. Keep track to ensure this happens.

How to Set Objectives with OMS

- Number of people with objectives agreed to
- Percentage of people with objective set
- Number of one minute praisings
- Number of one minute reprimands; ratio, etc.

Measurement

Completion to goal.

Reward

Performance Appraisal System; daily numbers game.

STRESS ASSESSMENT TECHNIQUE

You can assist your staff tremendously by paying attention to them and putting your focus on stressors they are facing on the job. You can also assist them by focusing on whether or not they are exhibiting any of the tell-tale warning signs that show they are heading toward stress-related illnesses.

The Stress Assessment Technique (SAT) utilizes a number of different assessment approaches in evaluating an individual's likelihood of becoming ill from facing stress. Use the following steps in looking at the individuals that work for you:
1. Check for the warning signs. They are:
    - General irritability, hyperexcitation, or depression

- Pounding of the heart
- Dryness of throat and mouth
- Impulsive behavior, emotional instability
- Overpowering urge to cry or run and hide
- Inability to concentrate
- Feelings of unreality, weakness, dizziness
- Predilection to becoming fatigued
- Floating anxiety
- Tension and alertness
- Trembling or nervous ticks
- Tendency to be easily startled
- High pitched nervous laughter
- Stuttering or other speech difficulties
- Bruxism (grinding teeth together)
- Insomnia
- Hypermotility, moving for no reason, cannot relax
- Sweating
- Frequent need to urinate
- Diarrhea, indigestion, queasiness, vomiting
- Migraine headaches
- Premenstrual tension or missed cycles
- Pain in neck and lower back
- Loss of, or excessive, appetite
- Increased smoking
- Increased use of drugs or alcohol
- Drug or alcohol addiction
- Nightmares
- Neurotic behavior
- Psychosis
- Accident proneness

2. Are your co-workers in any of the following "high stress" job categories?
    - Laborer
    - Secretary
    - Inspector
    - Lab technician
    - Office manager

- Foreman
- Manager
- Waitress
- Machine operator
- Farm owner
- Mine operator
- Printer

3. Are they expressing any stressors on these categories?
   - Job satisfaction
   - Physical conditions
   - Organizational factors
   - Work load
   - Work hours
   - Work task
   - Career development
   - Downsizing
   - Big Brother syndrome
   - Takeovers
   - Too much responsibility
   - The "ostrich syndrome"

4. If agreed, have your co-workers complete the nine self-assessment exercises in Appendix B. Be available if they wish to talk about the results on the profile summary.

5. Assess your company by filling out the profile questionnaire, Exhibit 11-2, which lists some of the key stressors in organizations.

6. Following are typical strategies that reduce job stress. How many are you using now?
   - *Job redesign*, modifying the content of work, enriching the tasks done, or rotation;
   - *Organizational modification*, giving greater autonomy, more ownership, more delegation of decision making and problem solving to the worker;
   - *Ergonomic redesign*, using what we know to make the job user friendly;
   - *Modifying the working space and time*, removing crowding or isolation, allowing rest periods;
   - *Reducing forced overtime* through better manpower planning and better scheduling;

AREAS	PROBLEMS/CONSIDERATIONS
COMPANY_____ DATE_____	
1. Have you had a climate/stress survey of any type?	
2. What problems were identified?	
3. Are you/have you been in a merger or takeover?	
4. Are you downsizing?	
5. Are you losing your employee feelings of loyalty?	
6. Do you have an ongoing job enrichment program in place?	
7. Where (at what level) are most decisions made?	
8. Do you have an ergonomic analysis program in effect?	
9. Do your people work overtime? Is it forced on them?	
10. Is there an effective effort to increase employee participation?	
11. Do you use MBO?	
12. Are you experimenting with autonomous work groups?	
13. Do you have an active wellness group?	
14. How did you score on the Likert Climate Scale?	
15. Are your people treated as mature individuals?	
16. How does your organization deal with conflict?	

Exhibit 11-2. Stress Assessment Technique company profile questionnaire.

	COPE WITH STRESS			REDUCE STRESS		
	Now?	Should?		Now?	Should?
Use stress audit?	☐	☐	Climate survey?	☐	☐
Awareness programs?	☐	☐	Been in merger?	☐	☐
Offer training in			Downsizing?	☐	☐
Goal alternative?	☐	☐	Losing loyalty?	☐	☐
Time management?	☐	☐	Job enrichment?	☐	☐
Delegation?	☐	☐	Level of decisions?	☐	☐
Nutrition?	☐	☐	Ergonomic analysis?	☐	☐
Exercise?	☐	☐	Forced overtime?	☐	☐
Assertiveness?	☐	☐	Participation?	☐	☐
Goal path?	☐	☐	MBO?	☐	☐
Thought-stopping?	☐	☐	Autonomous groups?	☐	☐
Relaxation recall?	☐	☐	Wellness programs?	☐	☐
Relaxation response?	☐	☐	Treat people maturely?	☐	☐
PRT?	☐	☐	Conflict strategies?	☐	☐
Meditation?	☐	☐			
Planning?	☐	☐			
Biorhythms?	☐	☐			
Smoking cessation?	☐	☐			
Form support groups?	☐	☐			
EAPs?	☐	☐			

Exhibit 11-3. What is your company doing now or what should it be doing to help the individual on the job?

- *Providing more information* on everything, so the workers can feel they are a part of the organization;
- *Allowing worker input* before changes are made in the work set-up;
- *Allowing participation* in most decision making.

7. In Exhibit 11-3 are typical strategies that companies use to reduce stress and help people cope. How many of these strategies are in place at your company?
8. From all of the above, write an objective on how you can help those under your supervision or within your group.

How to Set Objectives with the SAT

Number of people assessed; percentage of people assessed.

Measurement

Completion to goal.

Reward

Performance appraisal system; daily numbers game.

CHAPTER 12 Safety Techniques That Work

Motivating

WORKER SAFETY ANALYSIS

SAFETY PROFESSIONALS AND SUPERVISORS are generally aware of the concept of job-safety analysis: the systematic analysis of specific jobs to spot situations with accident potential. Worker Safety Analysis (WSA) is the systematic analysis of a worker. One tool that we can use in worker safety analysis is shown in Exhibit 12-1. Management can devise a form to assist the supervisor to look systematically at each worker to determine whether he or she is highly likely to make human errors. Are there logical reasons why any particular employee is likely to make such errors? Worker safety analysis can uncover these reasons.

Some of the subjects covered on the form in Exhibit 12-1 are optional, for example, information about biorhythms and LCUs. These items can be left off a form if the information is not available, although LCU information might be available to a supervisor who knows his or her people. Personality type and accident risk are also optional, and might not be filled in by many. These items maybe less valuable, from a safety standpoint, than the other items on the form. The value analysis is quite arbitrarily decided by the supervisor, and might or might not be useful.

Current motivational analysis is the most valuable item on this form. The supervisor should look at each item listed and determine what motivational pull it will have on the employee. Included are most of the important determinants of employee performance we have discussed in this book.

WORKER SAFETY ANALYSIS

Name_____ Date _____

Long-term analysis
   Biorhythmic Information (dates to watch): _____
   LCU Information; approximate units accumulated now: _____

Personality and value analysis:
   Personality type _____
   Accident risk _____

	Key Importance	No Importance
Value of work		
Value of safety		

Current motivational analysis _____ Turn ons _____ Turn offs
   Peer group_____
   Me (boss relations) _____
   Company policy _____
   Self (personality) _____
   Climate_____
   Job-motivation factors _____
      Achievement_____
      Responsibility _____
      Advancement _____
      Growth _____
      Promotion _____
      Job _____
   Participation _____
   Involvement _____

Current job assignment_____ High _____ Low
   Pressure involved _____
   Worry or stress
   Information processing need _____
   Hazards faced _____
   Other _____

Force-field analysis:

  Pulls to safety

  ↑
  ↓

  Pulls away from safety

Current assessment:
❏ OK   ❏ Discuss with worker   ❏ Training   ❏ Crisis intervention
                 ❏ Contract   ❏ Behavior modification
❏ Crisis intervention

Exhibit 12-1. Supervisor's Worker Safety Analysis form.

The supervisor should consider current job assignment in terms of the load it places on the person. The last item on the form is force-field analysis; the supervisor may choose to perform a small force-field analysis to determine the current pulls on the employee. The entire worker safety analysis might lead to some disposition, as shown at the bottom of the form.

Obviously, the supervisor who fills out this kind of form will need considerable training in order to understand the concepts involved. The purpose of worker safety analysis is to help the worker; its intent is to spot the causes of human error before an accident can occur.

Dealing with Human-Error Causes

The bottom of the form offers several dispositions. If no real problem is unearthed and no action is indicated, the "OK" box is checked. Other disposition options are: to discuss the analysis with the person, as in the case of a relatively minor problem; to send him or her to receive additional training, or to give it yourself; to administer crisis intervention, in the event that the analysis reveals something crucial and critical that must be dealt with; to draw up a contract for behavioral change; or to use behavior modification. Some of these dispositions will usually require additional comment.

Training

This is one of the simplest solutions, and usually one of the most ineffective. It assumes that we have identified a lack of knowledge or skill. If this assumption is correct, training is the proper solution and will be effective. If, however, the assumption is not correct, that is, the problem is not a lack or knowledge or skill, a different solution is indicated.

Crisis Intervention

Obviously, crisis intervention is indicated only if a severe, immediate problem exists.

Behavior Modification

Behavior modification is not new, and it is relatively simple to understand. The basic process involves systematically reinforcing positive behavior while ignoring or penalizing unwanted behavior. The end result is the creation of a more acceptable response to a given situation. The technique concentrates on a person's observable behavior and not on its underlying causes.

Contracts

A contract can also affect behavior change in an employee. A contract is an agreement to do something about something; an agreement between the supervisor and the employee. Contracts can be established to change behavior, to change feelings, or to change physical conditions such as high blood pressure and obesity. In a work situation, contracts are primarily used for the purpose of changing behavior. According to the book *The OK Boss* by Muriel James,[1] there is a five-step process for making contracts:

1. The establishment of a goal. James suggests that the question, "What do I want that would enhance my job?" be asked by the employee.
2. The definition of what needs to be changed to achieve the goal. What would I need to change so that I can reach my goal?
3. The determination of how much the person is willing to do to achieve the goal. What would I be willing to do to make the change happen?
4. The determination of measurement and feedback needed to accomplish the change. How would others know when I have affected the change?
5. What pitfalls are there in the way? How might I sabotage or undermine myself so that I would not achieve my goals?

Muriel James states that each of the five points should be discussed when a contract is made. Written answers to each of the questions are also helpful.

The contract is an excellent management tool for dealing with employees. The employee is allowed to participate in what is going on, which is always preferable to authoritarian enforcement.

Procedure

Use the form to assess each employee. If you choose, use it only as a guideline to your thinking, not a check-off sheet and with no names. Assure over time that each employee is assessed by you.

How to Set Objectives with the WSA

Number of WSA's completed; percentage of people analyzed, etc.

Measurement

Completion to goal.

Reward

Performance Appraisal System; daily numbers game.

INVERSE PERFORMANCE APPRAISALS

Inverse Performance Appraisals (IPA) are a way that any supervisor can get accurate information as to his/her management effectiveness. This is done through systematically asking his/her people on a regular schedule what they think are his/her strengths and weaknesses when it comes to safety performance.

The IPA Procedure

Reproduce the form in Exhibit 12-2 so that each of the employees that report to you have a copy. Call a meeting of your staff/team and explain that you would like their honest opinion of the way they are being managed by you when it comes to safety. Pass out the form. Suggest that they select one person to collect them and tabulate the results for you. Assure that no names are asked for or should be used. Leave the group to complete the forms.

Report back to the group the results of the appraisal in another meeting. Use the results to set objectives on improvement.

How to Set Objectives with the IPA's

Number of times completed; percentage of employees involved; analysis of results; changes made, etc.

Measurement

Completion to goal; analysis of results and changes made; improvement overtime.

Reward

Performance Appraisal System, daily numbers game.

SAFETY IMPROVEMENT TEAMS (SIT's)

Employee participation and involvement is one of the best ways to provide an environment that is motivational. When people have a piece of the action, are involved, it positively affects their behavior.

Department_____

Note: Do not sign your name. Your boss will not see this sheet. He or she will receive a summary of responses from this department.

Consider your boss and how he or she performs compared to your expectations.

Does your boss: Better than I would expect   Worse than I would expect
                                   10   1

Know you? _____
Understand you? _____
Know what your needs are? _____
Write any comments here you wish: _____
_____
_____
_____

Back you? _____
Listen to you? _____
Talk to you? _____
Write any comments here you wish: _____
_____
_____
_____

All your input? _____
Ask for your ideas? _____
Use your ideas? _____
Write any comments here you wish: _____
_____
_____
_____

Remove any barriers in your way? _____
Have enough influence with his or her boss? _____
Have enough influence with other departments _____
Write any comments here you wish: _____
_____
_____
_____

Talk down to you? _____
Treat you as a child? _____
Treat you as a subordinate? _____
Write any comments here you wish: _____
_____
_____
_____

Exhibit 12-2. Inverse Performance standards form.

SIT Procedure

Call your people together and suggest you need their help in safety. Ask them to join with other workers of their choice to help solve problems in the department. Allow them to work with people of their choice and select problems of their choice. Provide them with any technical help they ask for. Ask for a monthly (a regular) progress report.

How to Set Objectives with SIT's

Number of teams; percentage of people involved; number of projects worked on; solved, etc.

Measurement

Completion to goal.

Reward

Performance Appraisal System; daily numbers game.

CLIMATE ANALYSIS

Climate Analysis (CA) is a technique to assess your staff/team's perception of your organization.

Asking employees can be very useful in assessing the real corporate climate—assuming they will level with you. Whether or not this happens is probably dependent on trust and spending enough time in the process to get to where they will level with you.

Surveys

The second option is the survey. With this approach a lot of data can be gathered quickly and it can be processed rapidly by computer. But the survey instrument is crucial. It must be constructed by people who know what they are doing and that takes time. Each question must be carefully constructed and validated by a professional test development person working with subject experts. Exhibit 12-3 is a short, quick, rating your employees can give you, however.

Procedure

First fill out the form on Exhibit 12-3. Ask each of your people to assess his climate, as you explain each of the factors. Once they understand each

Category	Organizational variables	Virtually none / 1,2,3 occasionally 4 / Mostly at top / Very little / Downward / With suspicion / Usually inaccurate / Not very well	Some / 4, some 3 / Top and middle / Relatively little / Mostly downward / Possibly with suspicion / Often inaccurate / Rather well	Substantial amount / 4, some 3 and 5 / Fairly general / Moderate amount / Down and up / With caution / Often accurate / Quite well	A great deal / 5, 4, based on group / At all levels / Great deal / Down, up, and sideways / With a receptive mind / Almost always accurate / Very well	Item no.
LEADERSHIP	How much confidence and trust is shown in subordinates?	Virtually none	Some	Substantial amount	A great deal	1
LEADERSHIP	How free do they feel to talk to superiors about job?	Not very free	Somewhat free	Quite free	Very free	2
LEADERSHIP	How often are subordinate's ideas sought and used constructively?	Seldom	Sometimes	Often	Very frequently	3
MOTIVATION	Is predominant use made of 1 fear, 2 threats, 3 punishment, 4 rewards, 5 involvement	1, 2, 3 occasionally 4	4, some 3	4, some 3 and 5	5, 4, based on group	4
MOTIVATION	Where is responsibility felt for achieving organization's goals?	Mostly at top	Top and middle	Fairly general	At all levels	5
MOTIVATION	How much cooperative teamwork exists?	Very little	Relatively little	Moderate amount	Great deal	6
COMMUNICATION	What is the usual direction of information flow?	Downward	Mostly downward	Down and up	Down, up, and sideways	7
COMMUNICATION	How is downward communication accepted?	With suspicion	Possibly with suspicion	With caution	With a receptive mind	8
COMMUNICATION	How accurate is upward communication?	Usually inaccurate	Often inaccurate	Often accurate	Almost always accurate	9
COMMUNICATION	How well do superiors know problems faced by subordinates?	Not very well	Rather well	Quite well	Very well	10
DECISIONS	At what level are decisions made?	Mostly at top	Policy at top, some delegation	Broad policy at top, more delegation	Throughout but well integrated	11
DECISIONS	Are subordinates involved in decisions related to their work?	Almost never	Occasionally consulted	Generally consulted	Fully involved	12
GOALS	What does decision-making process contribute to motivation?	Not very much	Relatively little	Some contribution	Substantial contribution	13
GOALS	How are organizational goals established?	Orders issued	Orders, some comments invited	After discussion, by orders	By group action (except in crisis)	14
GOALS	How much covert resistance to goals is present?	Strong resistance	Moderate resistance	Some resistance at times	Little or none	15
CONTROL	How concentrated are review and control functions?	Very highly at top	Quite highly at top	Moderate delegation to lower levels	Widely shared	16
CONTROL	Is there an internal organization resisting the formal one?	Yes	Usually	Sometimes	No—same goals as formal	17
CONTROL	What are cost, productivity, and other control data used for?	Policing, punishment	Reward and punishment	Reward, some self-guidance	Self-guidance, problem-solving	18

Exhibit 12-3. Organizational profile.

factor, ask them to assess again quarterly. Work on the weak spots between each assessment for the areas you have some control over.

How to Set Objectives with CA's

- Number of times CA's completed;
- Number of CA's;
- Percentage of people involved;
- Analysis of results;
- Changes from CA's, etc.

Measurement

Measure completion of goal; changes made from results.

Reward

Performance Appraisal System; daily numbers game.

Notes

1. Muriel James, *The OK Boss* (Reading, Mass: Addison-Wesley Publishing Co., 1975).

APPENDIX A Supervisory Self-Appraisal

WHAT MAKES A GOOD MANAGER OR SUPERVISOR? Earlier we said there are no innate traits or abilities that do this. And this is true. However, we believe there are certain challenges a supervisor must face that will put him to the test. How he reacts in these situations determines whether or not he is a good manager. What are these situations? Here are some:

— *Change.* How does he react to change? The only thing you can be sure of in a supervisory position today is that you will have to face change.

— *Handling people.* Some supervisors are usually directive in approach (autocratic); some are democratic.

— *Aggression.* How does he handle situations in which he feels aggressive? What situations bring on aggressiveness?

— *Frustration.* Every supervisor faces this. How does he handle it?

— *Decision making.* Every supervisor must make choices. How does he react in the decision-making process?

— *Delegation.* How does he use his authority to get things done?

— *Interpersonal relations.* What personal qualities does he admire in other managers? Does he try to emulate these? Should he? What qualities does he detest? Does this limit his effectiveness?

— *Tolerance for ambiguity.* The hallmark of the managerial task is that it is often unstructured. How does he handle ambiguous responsibilities?

These are some, obviously not all, of the challenges that influence managerial effectiveness. In this section we ask you to look at yourself and how you handle or react to each of them.

It is an appraisal you do of yourself. Its intent is to have you look at yourself as you may never have looked at yourself before. There are no right or wrong answers. There is no scoring, no comparison of your score to others. Just read the section through, answer or respond as asked, then look at your responses; see if they tell you anything about yourself as a supervisor.[1]

YOUR RESPONSE TO CHANGE

In this first part of the program we ask you to answer yes or no, depending on which of these two choices is closer to your own understanding of yourself.

YES   NO

a. Do you dislike getting new and/or additional assignments?
b. Do you feel that experience is a poor teacher?
c. Do you find yourself making the same mistakes over and over?
d. Are you always up to your neck in today's work when tomorrow's work hits your desk?
e. Are you usually the last to give up on a cause?
f. Are people around you always doing the unexpected?
g. Do you usually come up with one answer for a problem rather than many possible solutions?
h. Do you find a routine more congenial than doing a constant variety of things?
i. Do situations often arise that surprise you?
j. Do you mistrust situations that involve a risk?

Total ____ ____

As you probably guessed, the questions focus on your reaction to changes in the people, work, and situations around you. If you answered yes to seven or more of the questions, you probably like things to go along at an even keel. You feel most comfortable with familiar people, situations, and tasks.

If you answered no to seven or more of the questions, you like variety in the situations, tasks, and people that you deal with. You expect things to change and welcome new experiences. But it is only a question of degree. Even if you answered all the questions yes, you are only relatively set in your ways. If you answered no to all the questions, you are simply more flexible than most people.

Either way, it is useful for you to know whether you are relatively flexible or relatively set in your ways. Knowing this will help you understand your reactions to many situations. It will help you prepare yourself for any situations or encounters that you can foresee. In looking at others, you should size them up as relatively rigid or relatively easygoing. When dealing with them, you will take this factor into consideration.

This knowledge about yourself and others is particularly handy in conflicts. Suppose that you are shorthanded. everyone under you is pulling his weight and more, but you know that you are not going to meet your quota for the next few months. You need more workers. Your superior thinks you have enough. Of course you'll try to change his mind. But when you know him or her (that is, when you know how rigid he or she is), you will know when to quit fighting for more people and make do with the personnel you now have.

FOUR BASIC MANAGERIAL ROLES

Answer the following questions as well as possible but spend no more than three minutes on each. If you cannot think of a situation like the one described, go on to the next. The situation does not have to be in a business context. Family, club, or school situations are good enough.

Outline briefly the last situation you took charge of, on your own—a situation in which you railroaded a job through to completion or where you completely dominated a meeting. Include anything you can remember that signaled you to adopt this autocratic role. Were you forceful enough to get the job done?

Outline briefly the last situation in which you had to keep still—a situation in which doing nothing or saying nothing was the only way to get the job done right.

Include anything you can remember that signaled you to adopt this self-effacing role. Were you able to keep quiet?

Outline briefly the last situation in which you had to be a democratic group leader. What made it important that everyone speak his mind? Were you able to coordinate the freedom of action or expression of the others in your group?

Outline briefly the last time you and someone else, or two other people, were unable to reach a meeting of the minds until you were able to figure out a compromise. Be sure to include any methods that you can remember using to help one or both parties to accept the compromise.

If you were unable to think of a situation to fit one of the models, don't be disappointed in yourself. Had you been given more time, you would have thought of an example for each. We all have to adopt different roles in different situations, or we would be constantly clashing with reality. As a manager, however, you will find that in order to reach all your objectives, you will need to assume these different roles consciously. No one finds this easy. Managers who are more flexible than most enjoy the compromising or the democratic role. Managers who tend to be set in their ways feel more comfortable with the autocratic or the self-effacing role. But all managers will meet situations that call for roles they find distasteful.

A managerial or supervisory career can cover a period of many years. Very often only long experience in uncongenial roles will make a manager fully effective in these roles. But before he can choose whether to be autocratic or democratic, he must have a feeling for the situation. Very few people can sense instinctively what role would be most effective in any situation. Nearly every manager must develop the ability to read situations either before he gets involved in them or while they are developing.

You must ask yourself which role you habitually adopt in situations. When you recognize in yourself the tendency to take charge or to keep quiet, to get everybody to agree or to compromise, you will know what comes easiest for you. You will be wary, therefore, of adopting that role in every situation you meet. This understanding of yourself will help you to be more resourceful in dealing with people and tasks. You won't have to change your whole personality, but you can change your manner of dealing with different situations.

YOUR AGGRESSIVE DRIVE

What kinds of friendly combat do you engage in? Every healthy person needs some form of socially acceptable warfare. No one is so thoroughly meek and mild that he has no need to express aggressive instincts. Put a check after any of the following outlets you use to channel your aggressive impulses. Don't check those you use only occasionally.

__Competitive sports (golf, handball, tennis)
__Arguments
__Cards, chess, or other competitive games
__Hard physical exercise
__A hobby that involves physical labor
__Wrestling with your children
__Kidding or teasing people
__Being embarrassingly honest with people
__Gossiping
__Getting ahead in business
__Trying to control the behavior of others
__Bawling out other drivers, salesclerks, your children
__Burying your hostility under sugary words

Now that you've completed our list, try your hand at making one of your own. Put down other outlets you've found handy when your aggressive drive backs up inside you.

Looking at yourself in this perspective may have brought to your attention behavior that you never considered aggressive before. You may even have discovered ways that your aggressive drive sneaks out around your conscious guard to conduct various kinds of commando warfare. Recognition of aggressive drives is the first step toward controlling them. Of course, there are days when you feel out of sorts and not yourself. Knowing that you may have aggressive drives building up inside you, clamoring for release, will not automatically make you feel wonderful. But you must realize that aggressive instincts are not all bad. We owe ambition, our love of sports and games, and our courage in the face of real danger to our very necessary aggressive drives.

The important consideration here is whether your aggressive impulses get in the way of your managerial performance. Does your kidding, your anger, your sarcasm, your pestering, your competitive drive, or your gossiping cripple you as a manager? It may be that they are a big help to you in getting the work out. But it's worth a look. If one or more of these habits don't help to advance you and your group toward the objectives you've set, then you need to find other outlets. Keeping a house in repair, hunting, getting involved in church or civic affairs are only a few. There are hundreds of ways to channel this aggressive energy into rewarding and satisfying activities for you and others.

YOUR REACTION TO FRUSTRATION

Frustration and the aggressive drive are similar. Both build tension which seeks various outlets. Frustration, however, pops up unexpectedly. Your frustration level goes up or down whereas your aggressive drive is a more constant force. You can plan regular activities (golf, poker, long walks) to channel the relatively constant pressures of the aggressive drive. But we never know when something or someone will frustrate us.

Outline briefly the last occasion when you had to stand by, silent and powerless, when you were itching to get into the action. It doesn't have to be a business situation. Also, explain briefly what happened inside you as you stood by.

Outline briefly the last occasion on which you chose to stand by silently because you knew it would be best for the people involved to go it alone. It doesn't have to be a business situation. Explain what went on inside you this time.

The feelings you described are some of your normal symptoms of frustration. You may have felt the same symptoms when you were holding yourself back as you felt when you were powerless to help. Usually people have more than one way of reacting to frustration. Some of the most common outlets for frustration include: unusual outbursts of anger, increased irritability, fits of quarrelsomeness, lots of complaining, smoldering silence, drinking bouts, and helter-skelter activity.

You, as a supervisor, notice uncommon behavior in a subordinate. Let's say he is suddenly not talking. He comes in, does his work, and leaves. No greetings, no good-byes. You wonder whether you should do something about it. If the employee is just working off frustration, there is little you can do. If he is not bothersome to others, let him get it out of his system his own way.

If he cannot vent his feelings in his normal way, he will bottle them up. Headaches, insomnia, stomach aches, or rashes can result. People go to doctors with these complaints and find out there is nothing physiologically wrong with them. Some of these people know what is frustrating them, others do not. The latter have buried frustration so successfully that they find it very difficult to find out what is picking at them. They continue to be miserable until they find out what is frustrating them. When people know what is bothering them, they can usually find some less painful outlet for these feelings.

YOUR DECISION MAKING

Now we'll ask you to examine the nine steps in making decisions. You should recognize them all because, consciously or unconsciously, you follow this sequence as you decide on any complex problem. Check those steps that cause you the most concern, take the most time, or demand the most energy.

__ 1. Deciding which decision to tackle first.
__ 2. Defining the exact nature of the problem.
__ 3. Collecting information pertinent to the problem.
__ 4. Ascertaining the minimum objectives any decision will have to achieve.
__ 5. Deciding what is possible and what is impossible.
__ 6. Thinking up many alternatives.
__ 7. Checking the alternatives to see which ones achieve the minimum objectives.
__ 8. Picking the best of whatever alternatives survive the last step.
__ 9. Putting your decision into action.

Each step is a decision. And some of the steps call for several decisions. Each step forms a part of the rational process of decision making. None of them can be called the most important one. Step 6 will be very important to a person who can never think of more than one or two possible solutions to any problem. He could get help on this step by calling a brainstorming conference. To another person, collecting information could be a big step because he dislikes research. A supervisor can often delegate research.

SUPERVISORS MUST DELEGATE

One of the benefits of being a manager or supervisor is that you can delegate many of your duties to subordinates. To get the most mileage out of your best abilities, you must manage your work and your people so that you yourself do those things which you do best.

You see yourself as responsible to your superiors for certain results. Someone can help you translate these results into concrete objectives, but you make the decision on what part each group member takes in achieving the objectives. Whenever possible, each member should do those tasks that he does best. But this is not always possible. You, being the most important member of the group, should parcel out the work so that you can concentrate as much as possible on what you do best. Some matters, like important decisions, cannot be delegated. There may be other matters that you may not handle well but which no one else can handle at all. Your time and energy are more valuable than any other person's in the group. It's your duty to make the most of the time and energy you have.

To determine how you work within several different areas related to managerial performance, fill out the appraisal chart on the next page. Try to generalize from your past experience when deciding where to put your check marks. Be frank with yourself when judging your abilities and feelings about these aspects of a supervisor's job.

The appraisal chart will be useful the next time you reorganize your group to meet its objectives. By referring to the chart, you'll be able to spot quickly those activities that you should try to delegate, either in full or in part.

Start with the column headed "I do it poorly." Consider carefully each activity you checked in this column. When you have decided on the feasibility of farming out as much as possible of that work, move to the next column and work your way over to the last column on the far right. By looking at your tasks in this way, you will be able to delegate those jobs that take too much of your time and energy and that you don't really have to do yourself. You can't delegate jobs just because they are unpleasant or

Self-Appraisal Chart

	I do it poorly.	I don't like to do it.	I find it very boring.	I find it very time-consuming.	I find it satisfying.	I do it very well.
Dealing with people						
Dealing with paper or forms						
Making small decisions						
Working with numbers or data						
Formulating plans or ideas						

difficult for you. But you will find that some of these jobs can be delegated, or that at least parts of them can be done by subordinates.

The next step is to parcel out the tasks and train your subordinates to do them. Each task will get more attention from a subordinate than you can give it. And in many cases, the subordinate will like the task more than you do.

To make the chart even more useful, you could add areas of your job that are not included in the chart. This would make the chart more complete and more specifically tailored to your situation.

PERSONAL QUALITIES YOU ADMIRE

Since entering the business world, you've met and worked with some impressive men and women. Perhaps you have studied a few of them and noted the distinctive ways in which they work, act, and express themselves. You may be trying to develop in yourself some of their habits and skills. Following the example of another person is the most powerful and effective means for developing your abilities.

You must be careful, however, whom you choose as your model. The skills you admire may be far afield from your own. Ignoring your own talents in order to pursue a goal not suited to you can be frustrating. It will surely diminish the impact of your unique contribution as a manager.

Speaking for yourself as a supervisor, which kinds of business professionals described below do you admire most? Put a check mark by only one or two of the choices.

___ A person who gets along exceptionally well with people.

___ A person who builds tight, productive organizations.

___ A person who is creative, an idea man.

___ A person who has wide knowledge, an answer for everything.

___ A person who expresses himself clearly and convincingly.

___ A person who is cool under fire, very self-controlled.

It is important to keep in mind that you especially admire the qualities you checked in your associates. If you had a supervisor that you thought highly of because he built a high-producing sales unit, you might not be very sympathetic to complaints that he was too overbearing and often lost his temper. You must be wary of overvaluing certain people simply because they possess qualities you admire.

A far greater danger, however, exists in undervaluing people who do not have the qualities you most admire. Look again at the choice you made. Think of the opposite of each quality you checked. For example, if you checked "a person who expresses himself clearly and convincingly," you would turn that around to its opposite: "A person who has trouble getting his ideas across to others."

When you have found the opposites of the qualities you admire, you will likely have the qualities that you most dislike in people you work with. It will be hard for you to see the strong points of people who have what you think of as negative qualities. You must, therefore, be on your guard that these negative qualities do not prejudice you against those who have them.

It is very hard to be objective about people. But a supervisor must strive to keep in perspective the strengths and weaknesses of each member of his group. Only when he has a balanced view of each can he help each to contribute his or her best efforts.

SUPERVISORS NEED A TOLERANCE FOR AMBIGUITY

Write down as many of the projects and tasks you were involved in last week as you can remember. For the sake of brevity, use your own descriptive names or abbreviations for them. Don't spend more than three minutes in writing them down.

After each task, write down whichever of the following terms described most accurately the status of the project: not under way, just begun, less than half done, more than half done, almost done, completed, liable to be junked, urgently in need of repair, stalled, status indeterminate.

You probably have developed an impressive list of projects. And, within the limits of three minutes and less than a full page of paper, you haven't listed everything you're involved in. It might take quite a while to remember and write down every job you worked on last week and to indicate the status of each one.

You may have had great difficult in determining the status of some projects. As a manager, you would be even more out of touch with day-today accomplishments than you are as a supervisor. You would see progress only from week-to-week or month-to-month. You would be coordinating the efforts of many people working individually and in various combinations on perhaps a hundred jobs. From a managerial point of view, dozens of programs will be inching toward completion.

All these factors contribute to make the manager's or supervisor's environment less structured than the worker's. As hectic and demanding as the worker's job can be, it is cut and dried compared to the manager's or supervisor's job. The supervisor's work is only vaguely defined; he faces a variety of apparently unrelated decisions and situations; he must adopt different roles on different occasions. A supervisor's capacity to operate in this loosely structured environment is called his tolerance for ambiguity.

By exposure, you will develop an increasing tolerance for ambiguity that will help you cope with the many and varied demands of this less-structured environment.

CONCLUSION

You have just appraised yourself as a supervisor. Based on your experience, you have examined your mental and emotional makeup and how it will fit into the framework of demands that you will meet in any managerial position. In some sections of the self-appraisal, you were asked to measure yourself against certain standards. In most of the program, however, you explored your strong points.

The manager or supervisor who builds on his strengths will contribute his best to his job. You measured yourself against managerial standards only in those areas in which certain abilities are critical. For example, it is essential for a manager to know himself, his subordinates, and his operation as thoroughly as possible.

It is likely that this self-appraisal made you look at yourself in a new light. By asking yourself, "Am I making the fullest use of my abilities?" and "Am I contributing my best?", we hope you have gained some insight into the strengths and abilities that will make you an even more successful supervisor and manager.

Notes

1. This supervisory self-appraisal is based on a preliminary version of a programmed managerial self-appraisal course developed by Allstate Insurance Company for its management development programs. Used with permission.

APPENDIX B: Stress Tests

TESTING FOR PROPENSITY

IN RECENT YEARS a number of tests have been developed to help assess the individual's propensity to stress problems. Each is described with a brief description of what the test intends to do and how to score it. Later in the chapter is a summary sheet for all scores.

Each test is a self-assessment device, each self-scored. The reader is urged to honestly take each and record scores on the summary sheet which, when completed, will give a picture of the total propensity profile showing the current situation and the areas in which needed improvement might be made.

Below are listed events which occur in the process of living. Place a check in the left-hand column for each of those events that have happened to you during the *last 12 months.*

Life Event	Point Values
_____ Death of spouse	100
_____ Divorce	73
_____ Marital separation	65
_____ Jail term	63
_____ Death of close family member	63
_____ Personal injury or illness	53
_____ Marriage	50
_____ Fired from work	47
_____ Marital reconciliation	45
_____ Retirement	45
_____ Change in family member's health	44
_____ Pregnancy	40
_____ Sex difficulties	39
_____ Addition to family	39
_____ Business readjustment	38
_____ Change in financial status	38
_____ Death of close friend	37
_____ Change to different line of work	36
_____ Change in number of marital arguments	35
_____ Foreclosure of mortgage or loan	30
_____ Change in work responsibilities	29
_____ Son or daughter leaving home	29
_____ Trouble with in-laws	29
_____ Outstanding personal achievement	28
_____ Spouse begins or stops work	26
_____ Starting or finishing school	26
_____ Change in living conditions	25
_____ Revision of personal habits	24
_____ Trouble with boss	23
_____ Change in work hours, conditions	20
_____ Change in residence	20
_____ Change in schools	20
_____ Change in recreational habits	19
_____ Change in church activities	19
_____ Change in social activities	18
_____ Mortgage or loan	17
_____ Change in sleeping habits	16
_____ Change in number of family gatherings	15
_____ Change in eating habits	15
_____ Vacation	13
_____ Christmas season	12
_____ Minor violations of the law	11

Score: _____

After checking the items above, add up the point values for all of the items checked.

Self-assessment Exercise 1, on need to adapt.

Choose the most appropriate answer for each of the 10 statements below as it usually pertains to you. Place the letter of your response in the space to the left of the question.

____ 1. When I can't do something "my way," I simply adjust to do it the easiest way.
    (a) Almost always true   (b) Often true
    (c) Seldom true   (d) Almost never true

____ 2. I get "upset" when someone in front of me drives slowly.
    (a) Almost always true   (b) Often true
    (c) Seldom true   (d) Almost never true

____ 3. It bothers me when my plans are dependent upon the actions of others.
    (a) Almost always true   (b) Often true
    (c) Seldom true   (d) Almost never true

____ 4. Whenever possible, I tend to avoid large crowds.
    (a) Almost always true   (b) Often true
    (c) Seldom true   (d) Almost never true

____ 5. I am uncomfortable having to stand in long lines.
    (a) Almost always true   (b) Often true
    (c) Seldom true   (d) Almost never true

____ 6. Arguments upset me.
    (a) Almost always true   (b) Often true
    (c) Seldom true   (d) Almost never true

____ 7. When my plans don't "flow smoothly," I become anxious.
    (a) Almost always true   (b) Often true
    (c) Seldom true   (d) Almost never true

____ 8. I require a lot of room (space) to live and work in.
    (a) Almost always true   (b) Often true
    (c) Seldom true   (d) Almost never true

____ 9. When I am busy at some task, I hate to be disturbed.
    (a) Almost always true   (b) Often true
    (c) Seldom true   (d) Almost never true

____ 10. I believe that "all good things are worth waiting for."
    (a) Almost always true   (b) Often true
    (c) Seldom true   (d) Almost never true

Scoring: 1 and 10: a = 1, b = 2, c = 3, d = 4   Score: _____
         2 – 9: a = 4, b = 3, c = 2, d = 1

Self-assessment Exercise 2, on ability to cope with frustration.

Choose the most appropriate answer for each of the 10 statements below and place the letter of your response in the space to the left of the question.

How often do you:

_____ 1. Feel like you have to work overtime to complete your work?
    (a) Almost always   (b) Very often
    (c) Seldom   (d) Never

_____ 2. Feel like your boss (the organization) forces you to work overtime when you do not wish to?
    (a) Almost always   (b) Very often
    (c) Seldom   (d) Never

_____ 3. Feel like you need an assistant?
    (a) Almost always   (b) Very often
    (c) Seldom   (d) Never

_____ 4. Feel confused as to exactly what your role is?
    (a) Almost always   (b) Very often
    (c) Seldom   (d) Never

_____ 5. Feel discouraged because you can't get to everything that needs to be done?
    (a) Almost always   (b) Very often
    (c) Seldom   (d) Never

_____ 6. Feel depressed because there isn't time to do quality work?
    (a) Almost always   (b) Very often
    (c) Seldom   (d) Never

_____ 7. Feel the need to skip lunch or coffee breaks to finish tasks?
    (a) Almost always   (b) Very often
    (c) Seldom   (d) Never

_____ 8. Believe you do not have enough authority?
    (a) Almost always   (b) Very often
    (c) Seldom   (d) Never

_____ 9. Are being judged by things beyond your control?
    (a) Almost always   (b) Very often
    (c) Seldom   (d) Never

_____ 10. Feel that work that must be done at home interferes with the time needed at work?
    (a) Almost always   (b) Very often
    (c) Seldom   (d) Never

Scoring: a = 4, b = 3, c = 2, d = 1   Score: _____

Self-assessment Exercise 3, on perception of overload.

Indicate the most appropriate answer for each of the 10 statements in the space provided.

_____ 1. I need more social situations.
   (a) Almost always true
   (b) Often true
   (c) Seldom true
   (d) Almost never true

_____ 2. At times I feel very much alone.
   (a) Almost always true
   (b) Often true
   (c) Seldom true
   (d) Almost never true

_____ 3. I dislike traveling and would rather be home with my family.
   (a) Almost always true
   (b) Often true
   (c) Seldom true
   (d) Almost never true

_____ 4. Whenever I have a free minute I find something to read.
   (a) Almost always true
   (b) Often true
   (c) Seldom true
   (d) Almost never true

_____ 5. I find a job transfer and move very difficult.
   (a) Almost always true
   (b) Often true
   (c) Seldom true
   (d) Almost never true

_____ 6. I need a great deal of variety in my work; I get bored easily.
   (a) Almost always true
   (b) Often true
   (c) Seldom true
   (d) Almost never true

_____ 7. I detest standing in lines and waiting.
   (a) Almost always true
   (b) Often true
   (c) Seldom true
   (d) Almost never true

_____ 8. Listening to a lecture usually bores me.
   (a) Almost always true
   (b) Often true
   (c) Seldom true
   (d) Almost never true

_____ 9. I'm a pack rat and find it difficult to throw things away.
   (a) Almost always true
   (b) Often true
   (c) Seldom true
   (d) Almost never true

_____ 10. I dislike not being with others.
   (a) Almost always true
   (b) Often true
   (c) Seldom true
   (d) Almost never true

Scoring: a = 4, b = 3, c = 2, d = 1   Score: _____

Self-assessment Exercise 4, on perception of deprivation.

Answer the following questions; place the letter of your response in the space to the left.

_____ 1. How many cigarettes do you smoke daily?
    (a) None
    (b) Less than 1 pack
    (c) 1-2 packs
    (d) More than 2 packs

_____ 2. How many alcoholic drinks do you average in a day.
    (a) None
    (b) 1
    (c) 1-2
    (d) More than 2

_____ 3. How many cups of coffee do you drink each day?
    (a) 2 or less
    (b) 3-4
    (c) 5-6
    (d) 7 or more

_____ 4. How much exercise do you get daily?
    (a) More than 30 minutes
    (b) Less than 15 minutes
    (c) 15-30 minutes
    (d) None

_____ 5. How often do you snack between meals?
    (a) Never
    (b) Once a week
    (c) Once a day
    (d) Between each meal

_____ 6. How many of the following have you been treated for: ulcers, migraines, hypertension, hives, allergies, black-outs?
    (a) None
    (b) One
    (c) Two
    (d) More than two

_____ 7. How many of the following do you practice regularly: transcendental meditation, relaxation response, other meditation?
    (a) Three
    (b) Two
    (c) One
    (d) None

_____ 8. Are you overweight?
    (a) No
    (b) 5-10 pounds
    (c) 10-20 pounds
    (d) More than 20 pounds

_____ 9. Do you take vitamins daily?
    (a) Yes
    (b) No

_____ 10. How often do you travel on business?
    (a) Never
    (b) 1-2 trips per year
    (c) 1 trip per month
    (d) More than 1 trip per month

Scoring: a = 1, b = 2, c = 3, d = 4   Score: _____

Self-assessment Exercise 5, on health habits.

Calculate your exposure to noise using the following chart.

Example	Overall sound pressure level (dB re 0.0002 microbar)	Hours exposed daily	
	10 ×	_____	= _____
Studio for sound pictures	20 ×	_____	= _____
Soft whisper (5 feet)	30 ×	_____	= _____
Quiet office	40 ×	_____	= _____
Audiometric testing booth	×	_____	= _____
Average residence	50 ×	_____	= _____
Large office	×	_____	= _____
Conversational speech (3 feet)	60 ×	_____	= _____
Freight train (100 feet)	70 ×	_____	= _____
Average automobile (30 feet)	74 ×	_____	= _____
Very noisy restaurant	80 ×	_____	= _____
Average factory	×	_____	= _____
Subway	90 ×	_____	= _____
Printing press plant	×	_____	= _____
Looms in textile mill	100 ×	_____	= _____
Electric furnace area	×	_____	= _____
Woodworking	110 ×	_____	= _____
Casting shakeout area	×	_____	= _____
Hydraulic press	120 ×	_____	= _____
50 hp siren (100 feet)	×	_____	= _____
Jet plane	140 ×	_____	= _____
Rocket launching pad	180 ×	_____	= _____

Total noise level = _____

Divide by 8 = _____
(Average hourly level)

Self-assessment Exercise 6, to measure exposure to noise.

Choose the alternative that best summarizes how you generally behave, and place your answer in the space provided.

____ 1. When I face a difficult task, I try my best and will usually succeed.
    (a) Almost always true   (b) Often true
    (c) Seldom true   (d) Almost never true

____ 2. I am at ease when around members of the opposite sex.
    (a) Almost always true   (b) Often true
    (c) Seldom true   (d) Almost never true

____ 3. I feel that I have a lot going for me.
    (a) Almost always true   (b) Often true
    (c) Seldom true   (d) Almost never true

____ 4. I have a very high degree of confidence in my own abilities.
    (a) Almost always true   (b) Often true
    (c) Seldom true   (d) Almost never true

____ 5. I prefer to be in control of my own life as opposed to having someone else make decisions for me.
    (a) Almost always true   (b) Often true
    (c) Seldom true   (d) Almost never true

____ 6. I am comfortable and at ease around my superiors.
    (a) Almost always true   (b) Often true
    (c) Seldom true   (d) Almost never true

____ 7. I am often overly self-conscious or shy when among strangers.
    (a) Almost always true   (b) Often true
    (c) Seldom true   (d) Almost never true

____ 8. Whenever something goes wrong, I tend to blame myself.
    (a) Almost always true   (b) Often true
    (c) Seldom true   (d) Almost never true

____ 9. When I don't succeed, I tend to let it depress me more than I should.
    (a) Almost always true   (b) Often true
    (c) Seldom true   (d) Almost never true

____ 10. I often feel that I am beyond helping.
    (a) Almost always true   (b) Often true
    (c) Seldom true   (d) Almost never true

Scoring: 1 – 6: a = 1, b = 2, c = 3, d = 4   Score: _____
         7 – 10: a = 4, b = 3, c = 2, d = 1

Self-assessment Exercise 7, on perception of self.

Answer the following questions; place the letter of your response in the space to the left.

Do you:

_____ 1. Often have tense muscles?
    (a) Almost always true   (b) Often true
    (c) Seldom true   (d) Almost never true

_____ 2. Often experience dryness in the mouth?
    (a) Almost always true   (b) Often true
    (c) Seldom true   (d) Almost never true

_____ 3. Sweat profusely?
    (a) Almost always true   (b) Often true
    (c) Seldom true   (d) Almost never true

_____ 4. Have difficulty expressing yourself?
    (a) Almost always true   (b) Often true
    (c) Seldom true   (d) Almost never true

_____ 5. Try to solve all problems immediately?
    (a) Almost always true   (b) Often true
    (c) Seldom true   (d) Almost never true

_____ 6. Create crises when there are none?
    (a) Almost always true   (b) Often true
    (c) Seldom true   (d) Almost never true

_____ 7. Dwell on past crises?
    (a) Almost always true   (b) Often true
    (c) Seldom true   (d) Almost never true

_____ 8. Experience hand tremors?
    (a) Almost always true   (b) Often true
    (c) Seldom true   (d) Almost never true

_____ 9. Have a pounding heart occasionally?
    (a) Almost always true   (b) Often true
    (c) Seldom true   (d) Almost never true

_____ 10. Experience hives?
    (a) Almost always true   (b) Often true
    (c) Seldom true   (d) Almost never true

Scoring: a = 4, b = 3, c = 2, d = 1   Score: _____

Self-assessment Exercise 8, to measure anxious reactivity.

Answer the following questions.

Do you: Yes   No

1. Eat, talk and walk quickly?   ____   ____
2. Get easily bored? Tune people out?   ____   ____
3. Get impatient with slow people?   ____   ____
4. Feel guilty when relaxing?   ____   ____
5. Forget small details?   ____   ____
6. Usually speak rapidly?   ____   ____
7. Like to own things?   ____   ____
8. Generally lean forward on your chair?   ____   ____
9. Measure yourself by goals achieved?   ____   ____
10. Do everything at a rapid pace?   ____   ____

Scoring: yes = 4, no = 1   Score: _____

Self-assessment Exercise 9, to measure Type A behavior.

These nine self-assessment exercises provide a beginning point in assessing an individual's (or your) propensity to experience stress when confronted with problems.

Exercise 1 looks at the need to adapt.

Exercise 2 looks at behavior when faced with frustrations and at the ability to cope in these situations.

Exercise 3 measures the perception of overload.

Exercise 4 measures the perception of deprivation.

Exercise 5 measures health habits.

Exercise 6 looks at exposure to noise.

Exercise 7 measures perception of self.

Exercise 8 measures anxious reactivity.

Exercise 9 measures Type A behavior (discussed earlier).

The nine exercises provide the profile of your propensity to experience stress.

TESTING FOR PROPENSITY

Exercise:	1	2	3	4	5	6	7	8	9
							Self-	Anxious	Type A
	Adaptation	Frustration	Overload	Deprivation	Health	Noise	Perception	Reactivity	Behavior

	• 400	• 40	• 40	• 40	• 40	•	• 40	• 40	• 40
Scores Indicative of High Vulnerability to Stressors	• 350	• 35	• 35	• 35	• 35	• 105	• 35	• 35	• 35
	• 300	• 30	• 30	• 30	• 30	• 95	• 30	• 30	• 30
Moderate Vulnerability to Stressors	• 250	• 25	• 25	• 25	• 25	• 85	• 25	• 25	• 25
	• 200	• 20	• 20	• 20	• 20	• 75	• 20	• 20	• 20
Low Vulnerability to Stressors	• 150	• 15	• 15	• 15	• 15	• 65	• 15	• 15	• 15
	• 100	• 10	• 10	• 10	• 10	•	• 10	• 10	• 10

Personal stress profile summary sheet.

APPENDIX C Safety and Health Law

The following article by Dr. James Abrams, which appeared in the August 1997 issue of *Occupational Health and Safety* magazine, is an excellent overview of the various laws related to occupational safety and health.*

O.S.H.A.—Other Safety and Health Acts

> A survey of workplace law
> finds there are many traps
> for the unwary occupational health
> and safety professional

SAFETY AND HEALTH PRACTITIONERS occasionally overlook legal requirements outside of the Occupational Safety and Health Act. Yet many additional statutes, regulations, rules, common law, and human resource issues potentially affect the safety and health obligations of businesses. The matters treated here may operate independently of OSHA, duplicate OSHA's regulatory objectives, operate concurrently with OSHA, be preempted by OSHA, or have control (that is, "primacy") over OSHA's requirements.

These laws and principles may also provide a source of scientific authority in interpreting OSHA issues. Finally, they are often an interdisciplinary crossing point for other professionals, such as risk managers, human resource professionals, security, and legal counsel. Awareness of these laws helps to assure a comprehensive corporate safety and health program.

---

*Reprinted with permission, © 1997, Stevens Publishing Corp.

Americans With Disabilities Act--Rehabilitation Act of 1973

Safety and health practitioners were introduced to an entirely new form of regulation when the Americans With Disabilities Act was passed. Many health-related laws operate to protect workers from hazards independent of OSHA, including ERISA, workers' compensation, dual capacity negligence doctrines holding health care professionals personally liable, and the Family and Medical Leave Act. ADA issues of special interest to safety and health professionals include:

- *Threats to Safety.* The ADA contains provisions to permit an employer-initiated action to exclude individuals with a disability who present a direct threat to the health or safety of the disabled worker or to others. The ADA further specifies that the disability not be removable by "reasonable accommodation."

  How are employers to determine if the applicant or employee poses a direct threat? The act directs employers to examine: a) the duration of the risk; b) the nature and severity of the potential harm; c) the likelihood that the potential harm will occur; and d) the imminence of the potential harm (29 CFR Section 1630.2). Employers are to meet this standard by evaluating each case for "the probability and severity of the potential injury." The individual's relevant work and medical histories are to be taken into account as part of the analysis.

- *Disability-Related Attendance Issues.* What about absenteeism occasioned by an alleged disability? Employers may be able to demand regular attendance while concurrently satisfying the employer's accommodation obligation.

  When analyzing the issue of disability-related absence, concentrate on the employee's qualifications. Further, define full-time work, how the compensation for this position was calculated, and whether management contemplated a specific number of hours, days, or percentage of attendance. Undertake an analysis to determine whether a reasonable accommodation can be achieved. Undue hardship on the employer is often self-defining in connection with this stage. Manage the attendance/absenteeism issue as you concurrently attempt to accommodate. Analyze all state and federal laws that might serve as a legal excuse. Observe and uniformly enforce all applicable, internal attendance/absenteeism policies.

- *Medical Communications Policies.* This issue is unrelated to OSHA's regulatory requirements concerning the location of employee medical records, 29 CFR Section 1910.20, "Access to Employee Exposure and Medical Records."

  A medical communications policy is relatively easy to create and serves to convey to the employee population at large the employer's medical requirements pursuant to the ADA and the employee's obligations necessary to obtain its protection. The employer may choose to outline actions that are necessary to verify an alleged disability (necessary medical exams and documentation) and criteria to determine whether an employee is able to perform the essential functions of the job with or without reasonable accommodation.

- *Substance Abuse.* Current alcoholics and recovered drug addicts have protection. Unsettled are the questions of "current" and illegal use of prescription drugs. The U.S. Department of Transportation, the Nuclear Regulatory Commission, and Department of Defense maintain separate regulations concerning these questions. BNA Policy and Practice Series, "Personnel Management," 247:4-1 et seq. (State regulations governing drug testing, employee usage, and drug policies are listed at §581 in the BNA materials. See DOT Model Anti-Drug and Alcohol Misuse Prevention Plans.)

Child Labor

The U.S. Department of Labor maintains a number of "Hazardous Occupations Orders" for employees below the age of 18 in non-agricultural occupations. 29 CFR Part 570 Subpart E.

Individuals working in the following areas who are below the age of 18 are regarded to be undertaking hazardous activities that are detrimental to their health and well-being:

1. Manufacturing and storing explosives.
2. Motor vehicle driving and outside helper.
3. Coal mining.
4. Logging and sawmilling.
5. Exposure to radioactive substances.
6. Power-driven hoisting apparatus.
7. Operating power-driven metal-forming, punching, or shearing machine.
8. Mining (other than coal mining).
9. Slaughtering or meat packing, processing, or rendering.
10. Power-driven woodworking machines.
11. Power-driven bakery machines.
12. Power-driven paper products machines.
13. Manufacturing brick, tile, and kindred products.
14. Power-driven circular saws, band saws and guillotine shears.
15. Wrecking, demolition, and shipbreaking operations.
16. Roofing operations.
17. Excavation operations.

Consumer Product Safety Act

Under the Consumer Product Safety Act, Public Law No. 92-573, 86 Stat. 1207 (1972), 15 U.S.C. §2951-84 (1995), manufacturers, distributors, and retailers of consumer products are to notify the Consumer Product Safety Commission when information is obtained that "reasonably supports the conclusion" that a product fails to comply with a CPSC rule or the product contains defects that create substantial product hazards. "Defect" includes, but is not limited to, faults, flaws, irregularities in a product, design choices creating risk, inadequate warnings, and inadequate instructions. The law of products liability is looked to in interpreting the CPSC regulations.

This act and the law of products liability may have implications for employers who unwittingly become "manufacturers" by fabrication of various tools, devices, and equipment for a worker's or contractor's use. Additionally, an employer may be able to shift or offset workers' compensation liability through an employee's, independent contractor's, or its own suit against a manufacturer for a product defect responsible for injury.

The Department of Transportation

The regulation of driving, substance abuse, and hazardous materials has made DOT, which includes agencies known by the acronyms FAA, FHWA, FRA, FTA, and RSPA, a full partner in the area of safety and health.

With respect to substance abuse, private sector employers have sweeping obligations under 49 CFR Parts 199 and 40. As of January 1996, all employers, regardless of size, who are otherwise subject to DOT regulation must comply with these regulations.

Policing prohibited employee conduct, administering tests (pre-employment, post-accident, random, reasonable suspicion, return to duty, follow-up testing), and maintaining records are required by DOT. Its regulations preempt state regulations on the same subjects. With respect to the ADA and FMLA, DOT has asserted there are no conflicts between the DOT and these statutes.

Fair Labor Standards Act

There are few labor regulations older than the FLSA. Now almost 60 years old, it has become an antedated relic in the service and information industry succesor to the post-Depression period where it originated. One of the more confounding aspects of the FLSA relates to its disciplinary regulation., 29 CFR Section 541.118.

Employers must exercise caution when disciplining an FLSA-exempt employee for safety and health violations. Employers who suspend or dock the pay of an FLSA-exempt worker endanger the exemption claimed from overtime liability for the employee being disciplined and all other exempt workers who are similarly situated.

Part-day-absence pay docking in the private sector is to be avoided unless it is for "infractions of safety rules of major significance." The safer course is to utilize alternative disciplinary means.

The Food and Drug Administration

The FDA is not a regulatory agency with which safety and health professionals outside the food and drug industry come into contact. However, the FDA's authority should not be overlooked in areas where it may have jurisdiction in addition to agencies such as OSHA. For example, the Federal Food, Drug and Cosmetic Act, 21 U.S.C. 371(a), authorizes the FDA to issue substantive, binding regulations for the efficient enforcement of the act. The act provides that a medical device is misbranded if its labeling is false or misleading in any particular. A misleading act may occur if labeling does not reveal facts that are material in light of the circumstances.

The FDA issued a proposed rule last year (Federal Register, Vol. 61, No. 122) to require warning labels on latex-containing medical devices. This includes latex gloves. Their use has increased dramatically because of concerns about bloodborne pathogens.

National Labor Relations Act

The NLRA, one of the oldest labor laws in the country, is traditionally regarded as applying to only unionized businesses. This law also applies to non-unionized settings when two or more employees have joined to protest, challenge, or oppose a term or condition of their work.

Of particular importance to the safety and health professional is 29 U.S.C. Section 143. It states, in pertinent part: "...nor shall the quitting of labor by an employee or employees in good faith because of abnormally dangerous conditions for work at the place of employment of such employee or employees be deemed a strike under this chapter."

Many work refusals due to unsafe conditions have indeed been protected under the NLRA for a variety of workplace practices.

Title VII

The Civil Rights Acts of 1964, as amended, and 1991 extend protection to various classes of workers. In doing so, the act also protects workers who experience discriminatory treatment in their work assignments--even when such assignments are intended to protect the worker.

For example, suppose a woman of childbearing age, a pregnant woman, or a man is working in an area that exposes her or him to a reproductive risk. What steps should an employer take and how should such steps be taken? Exclusionary actions by an employer create a risk independent of the risk the employer seeks to control, that is, the risk of a charge of gender-based discrimination. What about a fetus exposed to industrial hazards? What rights does a fetus have independent of a parent? When born, do these rights change? An employer may face civil liability, independent of workers' compensation, for injuries sustained by a child because of its mother's or father's workplace exposure.

The case of **Auto Workers v. Johnson Controls Inc.,** 449 U.S. 187 (1991), created a legal standard holding that the employer could not rely on Title VII's bona fide occupational qualification defense for women working in areas contaminated with lead. The employer's cost of compliance, potential exposure to separate civil liability for injuries sustained by a post-exposure birth, or the safety of the fetus itself were irrelevant, according to the U.S. Supreme Court. The court basically held that a sex-specific exclusionary policy could not be upheld.

Three questions remain under *Johnson Controls:* Can compliance costs ever justify a bona fide occupational qualification defense? Is full disclosure to a woman sufficient to bar a subsequent civil action by a child for workplace exposure(s)? Can an unexposed, non-employee spouse enjoin an employer from making a work assignment to the exposed or pregnant spouse?

It should be noted that a number of courts have permitted a discriminatory employment policy that was designed to ensure the safety of others. For example, two U.S. Supreme Court decisions have separately held that it is permissible to restrict a job assignment based on age if age can reasonably be established to pose a safety risk or to restrict a female security guard from being exposed to situations that might endanger penitentiary inmates — apart from the potential dangers to which the worker might be exposed.

Employers who may "discriminate" against a particular class of employees must consider whether their policy is necessary to protect the safety interests of a third party with respect to the business as a whole and whether they are, have, or can apply their policy uniformly. Consistency and uniformity appear to be the keys in this area of the law. The EEOC's telephone number is 800-669-3362.

Toxic Substances Control Act

TSCA §8(c) and TSCA §8(e) cover allegations of significant adverse reactions to human health or the environment resulting from a particular chemical substance or mixture, or information (based on tangible evidence) indicating that a particular chemical substance or mixture presents a substantial risk to health or the environment.

An allegation is a statement made "without formal proof with regard to evidence." The information must "reasonably support" the conclusion that the chemical substance or mixture presents a substantial risk. The source can be anyone, including employees, neighbors, customers, doctors, nurses, delivery truck drivers, plaintiffs in lawsuits, etc., as well as designed, controlled *in vivo* or *in vitro* toxicity studies; medical and health surveys; and ecological, epidemiological, biomonitoring, and bioaccumulation studies.

Also, evidence of effects in consumers, workers, or the environment resulting from a spill or contamination.

Obligations imposed by this act are recordkeeping (unless the effect is already known) and an immediate report to EPA (unless the information is already known). The act's stated purpose is to create a historical record that will provide a means to identify previously unknown adverse effects and to ensure that new information indicating substantial risk receives prompt attention from EPA.

Besides manufacturers, those who must comply are processors and distributors whose companies also manufacture.

Toxic Torts

Gasoline and chemical leaks; insecticide, fungicide, rodenticide applications; and a host of other industrial and non-industrial releases and emissions have created the area known as "toxic tort" litigation. The liability theories arising under toxic tort litigation include nuisance, strict liability for defective products or ultrahazardous and abnormally dangerous activities, negligence, trespass, and intentional tortious acts such as assault and battery.

Missouri has enunciated one often-cited test in *Elam v. Alcolac, Inc.*, 765 S.W. 2d 42 (Mo. Ct. App. 1988): cert. denied, 110 S. Ct. 1989. A substantial factor standard of proof was articulated. In deciding whether a defendant's conduct is a substantial factor for causing harm, the court indicated the following questions should be asked and answered: Was there enough exposure to an identified harmful substance to activate disease? Is there a demonstrable relationship between the substance and the biological disease? Has the plaintiff been diagnosed with disease? Is there an expert opinion that the plaintiff's condition correlates with exposure to the harmful substance? Is the defendant responsible for the etiological agent which causes this disease?

Circumstantial evidence was permitted in the case.

Contingent Liability Under Various Labor Laws

The term "contingent workers" covers temporary workers; long-term temporary assignments; loaned servants; master-vendor arrangements; in-house temporary employees; payrolling/staffing; part-time employees; seasonal workers; casual workers; independent contractors; contract technical workers; employee leasing; outsourcing or managed services; temp-to-perm programs; temp-to-lease programs; and self-employed workers.

With approximately 6 million contingent workers providing services in a $40 billion industry, health and safety professionals need to be alert to the various tests as to who is an "employer" and "employee" under the Internal Revenue Code, the Fair Labor Standards Act, the National Labor Relations Act, workers' compensation, OSHA, ERISA Title VII, and other labor laws.

While the Council of Economic Advisers has taken the position that there is no standard definition for a "contingent worker," the use of other-than-regular and other-than-full-time employees is growing in all sectors of the country. When it comes to assessing liability for the improper use of a contingent worker, the agency or forum examining the issue will determine whether the individual asserting a claim is or is not an employee of the defendant.

Not all of the proceeding laws utilize the same test to determine who is an employee and who is an independent contractor or non-employee. Liability for employers is controlled by these tests.

Worker Involvement/Empowerment And Self-Directed Work Teams

Recently popularized, these groups have, in many instances, substituted, altered, or replaced management's judgement on subjects traditionally viewed as proprietary policy or executive subject matter, or matters of corporate governance.

The implications for safety-related liability are obvious. A product's manufacture implicates workers in decisions about its design, formerly reserved for management. A team decision affects a contractor's performance, resulting in injury and liability. A team decision, being the equivalent of a management decision, is made on a subject where personal liability could be imposed.

Corporate insurance and indemnification policies should be reviewed for appropriate coverage.

Alternative Dispute Resolution and Settlements

Employment litigation has increased by 400 percent in the past 20 years. Litigation in federal courts is estimated to be more than 10 percent of all filings. Some states report as much as half of their filings are employment-related. Facilitators, mediators, conciliators, fact finders, early neutral evaluators, and arbitrators are all new job titles on the menu of non-judicial dispute resolution alternatives.

Many federal agencies have implemented various ADR mechanisms. Check to see whether the agency with which you are working has such a mechanism.

In the resolution phase of any dispute, there are a number of caveats to be observed. A "settlement" typically takes the form of a contractual agreement between the disputants. Certain terms are essential for both sides and to conform to various legal doctrines. When a governmental agency action is or may be involved, exercise particular caution as to what you are agreeing to be settled. This may even be true when a settlement is effected between two or more private parties, and a government agency later seeks to discover and use the settlement's contents.

How a civil settlement with EPA, OSHA, or another government agency is worded can be critical if there is the potential for a future criminal action. How a civil subpoena or administrative warrant is treated may also be important to the corporation facing a potential criminal action.

Rule 408 of the Federal Rules of Evidence generally bars the admission of settlement discussions and compromises to prove the validity of a disputed claim. Unresolved is how these documents may enter into a potential criminal proceeding. The Second, Seventh and Ninth U.S. Circuit courts of Appeals have ruled that Rule 408 is inapplicable in criminal proceedings. Federal Rule of Evidence 801(d)(2)(C) or (D) could allow the government to introduce what you state in a civil dispute in a subsequent criminal proceeding.

If at all possible, affirmatively deny wrongdoing and attempt to have the government agree that it will not use your civil settlement in any subsequent criminal proceeding. The same precautions should be followed with state agencies.

Black Holes for Safety and Health Professionals

Finally, many, many legal traps exist for the unwary health and safety professional. These are found in the laws discussed above, and in other laws and employment issues as well. Among them are:

- ADA—"physical or mental impairment substantially limiting major life activity."
- Attorney-client privilege and the work product doctrine.
- Environmental whistleblowing enforced by the U.S. Department of Labor.
- Equal Access to Justice Act, Rule 11 of the Federal Rules of Civil Procedure, and miscellaneous fee-shifting rules.
- Ethical and legal obligations of corporate ombudsmen receiving safety and health complaints.
- Evidentiary problems for health and safety professionals who participate in disciplinary actions.
- Expert witnesses and "junk science."
- FMLA—serious medical condition
- Internal appraisals, assessments, audits, evaluations, inspections, and investigations.
- "Knowing endangerment" under environmental laws.
- Local building and fire department life safety regulations.
- Motor vehicle accidents—where workers' comp ends and liability shifting, indemnification, and punitive damages begin.
- Personal liability of supervisors.
- Premises liability.
- Privileged investigations.
- Safety and health-related discipline. See 29 CFR Section 1977.6; *Brock v. L.E. Meyers Co.*, 818 F. 2d 1278 (6th Cir. 1987): Colorado Workers Compensation Statute (C.R.S. §8-42-112).
- Search-and-seizure and worker privacy--testing, monitoring, accessing medical information, retrieving, listening, viewing, observing, and confiscating.
- Sexual harassment.
- State health department regulations, labor codes, and licensing boards.
- Telecommuting.
- Workers' comp regulations concerning extrahazardous materials.
- Workplace violence.
- Workplace communication practices—signs, notices, posters, recordkeeping, certifications, training, warnings, and labeling.

James Abrams, Ph.D., is a senior attorney with Texaco in Denver, Colo. He formerly was a trial attorney with the Office of the Solicitor in the U.S. Department of Labor, and an industrial hygienist with OSHA. The views expressed in this article are those of the author, provided for general reference use. The information provided does not constitute legal advice.

Index

A
Abrams, James, Ph.D., 219
Accident(s), 134-39
    cause, investigating for, 141-56. *See also:* Incident Recall Technique (IRT); Safety Sampling (SS); Statistical Safety Control (SSC) *and* Technique of Operations Review (TOR)
    costs, 90, 92-93
    drug and alcohol problems causing, 137-38
    key factors in, 89
    non-reporting of, 153
    physical and mental problems causing, 136-39
    prevention, Safety Sampling (SS) technique for, 141-45
Accident causation
    Schulzinger's studies of, 135
    theories, 9-10, 86-87
Accident investigation, 83-90
    rating sheet, 88
Accountability
    fixing, 49-50
    and measurement, 57-59
    and roles, 43-45
    systems, 37-44
Acquisitions, corporate, and total incidence rates, 78
Activity measures, 58
Aggressive drive, 198-99
Alcohol consumption and accidents, 138
Ambiguity, assessing tolerance for, 204-205
Americans With Disabilities Act, 220-21, 227
Anxious reactivity, self assessment of, 215
Apathy, 26
Argyris, Chris, 26-27, 108, 112
Attendance, disability-related, 220
Attitude(s), 113-120
    development, 120
    employee and supervisor, compared, 115
    group influence on, 120-128
    and safety, 113-15
Authority, 52-55
Aversives and positives, 117-19

B
Behavior
    contracts, 188
    Type A, measuring, 216
Behavior-based safety, 16, 30-31
    measurements, 30

Behavior Observation System (BOS), 79

C
Change
    assessing response to, 195-96
    self-assessment, 208
Charting, 146-47
Child labor, 221
Climate Analysis (CA), 191-93
Coaching, 101-104, 175-84. *See also:* Job Safety Observation (JSO); One-Minute Safety System; One-on-One contacts; Safe Behavior Reinforcement *and* Stress Assessment Technique
Consequences, 116-17
Consumer Product Safety Act, 221-22
Contingent liability under various labor laws, 225
Contracts, 188
Controls, 107-108
    common, in safety programs, 12-13
Cresswell, Thomas J., 14
Culture companies, 16
Cumulative Trauma Disorders (CTD's), 169-70

D
Dalton, Melville, 48
Decision making, assessing, 200-201
Decision points, 131-34
Delegation of authority, 201, 203
Deming's 14 "Obligations of Management", 27-28
Department manager accountabilities, 42-43
Department of Transportation, 222
Deprivation, self-assessment of, 211
Discipline, 49, 128-34
    guidelines in using, 128-30
    safety and health-related, 227
Dispute resolution, 226-27
Domino Theory of Accident Causation, 9, 86
Drug problems and accidents, 137-38
Dual factor theory, 113

E
Earnest, Gene, 76
Employee centeredness, 25
Engineering, inspection, maintenance (EIM), 66
Equal Access to Justice Act, 227
Ergonomic Analysis (EA), 169-73
    checklist, 172
    overview, 171

F
Fair Labor Standards Act, 222
Feedback and learning, 103
First-level manager's weekly safety report, 41
First line supervisor
    accountabilities, 40-41
    roles, 43-45
Food and Drug Administration, 222-23
Foreman's inspection form, 99
Forms/checklists
    accident cost report, 92
    accident report, 84-85
    Climate Analysis organizational profile, 192
    Ergonomic Analysis, 172
    foreman's inspection, 99
    for identifying key facts, 91
    incident recall, 154
    inverse performance standards, 190
    investigation rating sheet, 88
    Job Analysis worksheet, 158
    Job Safety Observation, 177
    organizational profile, 192
    OSHA Check Off, 169
    report of accident investigation, 87
    safety inspection checklist, 96
    self-appraisal chart, 202
    Safety Sampling, 143
    Stress Assessment Technique questionnaire, 183
    summary of unsafe practices, 97-98
    TOR supervisor's incident investigation report, 151
    typical inspection, 95
    Worker Safety Analysis, 186
    worksheet for analyzing job safety, 160
Frustration
    ability to cope with, 209
    assessing reaction to, 199-200

G
Group(s)
    building strong, 124-28
    influence, 120-28
    norms, 121-22
    pressure, 122-24
    symbols, 126-27

H
Hannaford, Dr. Earle, 113-15
Hazard Hunt (HH), 161-63
Hazards, inspecting for, 157-74. *See also:* Ergonomic Analysis (EA); Hazard Hunt (HH); Job Safety Analysis (JSA) *and* OSHA Compliance Check (OCC)
Health habits, self assessment, 212
Heinrich, H.W., 9-11, 13, 23
Herzberg, Frederick, 25
    job enrichment theory, 112-13
Hygiene factors, 25

I
Incentives, 25
Incident rates, total, 77-81
    changes following recent acquisitions, 78
Incident Recall Technique (IRT), 152-55
Incongruency Theory, 108-109
Industrial hygiene and safety, key elements, 77
Inspecting for hazards, 93-101, 157-74
Internal company environments, 22-27
International Classification of Diseases (ICD), 89
Inverse Performance Appraisals, 189-90

J
James, Muriel, 188
Job
    enrichment, 25
    performance, and abilities, 35-36
Job Safety Analysis (JSA), 157-61
Job Safety Observation, 175-78
    worksheet, 177

K
Key elements of industrial hygiene and safety, measurement of, 76-81

L
Leaders, choosing, 48
Learning, 103, 113-15. *See also:* Training
    conditions affecting, 116
    by imitation (modeling), 119
Likert, Dr. Rensis, 25, 110

M
Mager, Robert, 115-16
Management, 110-13
    behavior-based, 16, 30-31
    delegation of authority by, 201, 203
    employee contracts, 188
    influence of, 104-105
    participation in safety programs, 66, 68-70
    roles, 197-98
Management Safety Obligations, 28
Medical communications policies, 220
McGregor, Douglas, 110
Measurement, 57-59
    of key elements of industrial hygiene and safety, 76-81
Menu System 40-41
Modeling, 119

Motivation, 49, 66, 104-107, 185-94. *See also:* Climate Analysis (CA); Inverse Performance Appraisals; Safety improvement teams *and* Worker Safety Analysis (WSA)
    attitudes affecting, 113-20
    and learning, 103
    positive, 117-19, 133, 179
Mueller, Paul, 148
Multiple causation, 86

N
National Council on Alcoholism, 138
National Labor Relations Act, 223
Noise, self-assessment measuring exposure to, 213

O
One-Minute Safety System, 180
One-on-One contacts, 178-79
OSHA, 9
    influence of, on safety programs, 12-13
    Safety Program Guidelines, 24
OSHA Compliance Check (OCC), 163-69

P
Performance model, 34-37
Personal stress profile sheet, 217
Personal qualities, 203
Personality traits of a good supervisor, 47-48
Physical and mental problems and accidents, 136
Pope, W.C., 14
Positive reinforcement, 179
Proctor & Gamble Company, 76
Product liability and worker teams, 226
Productivity of supervisors, 111
Punishment and safety, 130-34

R
Records, 10
Reinforcement and learning, 103
Repetitive Motion Injuries, 169
Results measurement, 58
Rewards, 22
    value of, 34-35
Role(s)
    and accountability, 43-45
    perception, 36
    in safety, 6

S
Safe Behavior Reinforcement, 179
Safety
    and attitudes, 113-15
    behavior-based, 16-17
    and group influences, 120-28
    inspection checklist, 96
    instruction, 113-20
    integration, 62-63
    and punishment, 130-34
    report, 41
    sampling technique (SS), 141-45
    threats to, 220
    "Three E's" of, 23
Safety by objectives (SBO), 38-40
Safety management
    accountability systems, 37-44
    decision points, 131-34
    performance model, 34-37
    principles of, 17
    tools, 22
Safety media, 11
Safety philosophy, 14-18
Safety program(s), 9-13, 59, 62-63, 65-70, 114
    areas, 67
    training in, 10
Safety Sampling (SS) technique, 141-45
    worksheet, 143
Sayles, Leonard, 69
Schleh, Edward C., 48
Schulzinger, Dr. Morris, 135
SCRAPE, 37-38
Search and seizure, 227
Self, self-assessment on perception of, 214
SPC tools, 145-47
Statistical safety control (SSC), 145-48
Stress
    and accidents, 139
    personal profile of, 217
    tests, 207-17
Stress Assessment Technique, 180-84
    profile questionnaire, 183
Substance abuse, 221-22
Summary of unsafe practices, 97-99
Supervisor(s)
    accident cost report, 92
    accident report form, 84-85
    accountabilities, 40, 49-50
    attitudes, 114-15
    authority, 55
    discipline, 128-34
    identification of accident-prone workers, 138
    influence of, 105-106
    job defined, 47-50
    non-reporting of accidents by, 153
    participation, 66, 69-71
    performance, factors affecting, 71
    personality traits of a good, 47-48
    productivity of, 111
    report of accident investigation, 87
    response to change, 196-97
    responsibilities, 50-52
    safety performance model, 34-37
    tolerance for ambiguity, 204-205

TOR incident investigation report, 151
traditional vs. nontraditional tasks of, 6
Supervision
    key elements of, 48-50
    sins of, 107-108
Symbols, developing group, 126-27
System safety, formula for excellence in, 18

T
29 CFR 1910, Subparts D-S and Z, 164-68
Taylor, Frederick W., 71
Team(s), 73-81. *See also:* Group(s)
    building stronger, 110-11, 125-28
    concept, 74
    and product liability, 226
    safety improvement, 189-91
Technique of Operations Review (TOR), 52, 148-52
    Cause Code, 148-50
Theory X vs. Theory Y supervisors, 110, 112
Three E's of Safety, 23
Title VII, Civil Rights Acts of 1964 and 1991, 223-24
TOR. *See:* Technique of Operations Review (TOR)
Total Quality Management, 27-31
Toxic Substances Control Act, 224
Toxic Torts, 224-25
Training, 10, 49, 66, 187. *See also:* Learning
    of foremen in positive motivation, 133-34

U
Unsafe practices, summary of, 97-98

W
Weaver, D.A., 148
Weeky safety report, 41
Wellness. *See:* Health habits
Work
    groups, building effective, 110-11
    overload, self-assessment of, 210
Worker
    attitudes, 113-20
    problem, 134-39
Worker Safety Analysis, 185-89
    supervisor's checklist, 186
World Health Organization, ICD coding, 89
Worker's Compensation, 31